Lecture Notes
in Business Information Processing

270

Series Editors

Wil M.P. van der Aalst
  Eindhoven Technical University, Eindhoven, The Netherlands
John Mylopoulos
  University of Trento, Trento, Italy
Michael Rosemann
  Queensland University of Technology, Brisbane, QLD, Australia
Michael J. Shaw
  University of Illinois, Urbana-Champaign, IL, USA
Clemens Szyperski
  Microsoft Research, Redmond, WA, USA

Jorge Munoz-Gama

Conformance Checking and Diagnosis in Process Mining

Comparing Observed and Modeled Processes

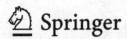 Springer

Jorge Munoz-Gama
Departamento de Ciencia de la Computación
Pontificia Universidad Católica de Chile,
  Escuela de Ingeniería
Macul, Santiago
Chile

ISSN 1865-1348   ISSN 1865-1356 (electronic)
Lecture Notes in Business Information Processing
ISBN 978-3-319-49450-0   ISBN 978-3-319-49451-7   (eBook)
DOI 10.1007/978-3-319-49451-7

Library of Congress Control Number: 2016957478

Printed on acid-free paper

This Springer imprint is published by Springer Nature
The registered company is Springer International Publishing AG
The registered company address is: Gewerbestrasse 11, 6330 Cham, Switzerland

To Loly, Emilio, Alex, and the rest of my family and friends.

Preface

This book encompasses a revised version of the PhD dissertation of Jorge Munoz-Gama written at the Computer Science Department of the Universitat Politècnica de Catalunya (Spain). In 2015, the dissertation won the "Best Process Mining Dissertation Award," assigned by the IEEE Task Force on Process Mining to the most outstanding PhD thesis, discussed between 2013 and 2014, focused on the area of business process intelligence.

In the past few decades, the capability of information systems to generate and record overwhelming amounts of event data has witnessed an exponential growth in several domains, and in particular in industrial scenarios. Devices connected to the Internet (Internet of Things), social interaction, mobile computing, and cloud computing provide new sources of event data and this trend will continue in the next decades. The omnipresence of large amounts of event data stored in logs is an important enabler for process mining, a novel discipline for addressing challenges related to business process management, process modeling, and business intelligence. Process mining techniques can be used to discover, analyze, and improve real processes, by extracting models from observed behavior. The capability of these models to represent the reality determines the quality of the results obtained from them.

The aim of this book is conformance checking, one of the main areas of process mining. In conformance checking, existing process models are compared with actual observations of the process in order to assess their quality. These models are typically the result of a hand-made analysis influenced by the bias of the analysts and the process owners, modeling a possibly outdated representation of the process. Conformance checking techniques are a way to visualize the differences between the assumed process represented in the model and the real process in the event log, pinpointing possible problems to be addressed, and the business process management results that rely on these models.

Conformance checking is a complex multidimensional analysis, where orthogonal dimensions such as fitness (measuring and ensuring that models capture all the behavior in the log) and precision (not including unnecessary behavior) determine the quality of the models. Moreover, a conformance analysis of real-life processes may overcome additional challenges such as the presence of noise or the size of the models.

The first part of the book focuses on analyzing and measuring the precision dimension of conformance, where models describing precisely the reality are preferred to overly general models. The book includes a novel technique based on detecting *escaping arcs*, i.e., points where the modeled behavior deviates from the one in the log. The detected escaping arcs are used to determine the precision between log and model, and to locate possible actuation points in order to achieve a more precise model. The book also presents a confidence interval on the provided precision metric, and a multi-factor measure to assess the severity of the detected imprecisions. These techniques open the door to noise-robust analysis of real-life processes and the possibility of ranking the misconformances detected regarding the importance and potential impact in the process.

Checking conformance can be time consuming for real-life scenarios, and understanding the reasons behind the conformance mismatches can be an effort-demanding task. The second part of the book changes the focus from the precision dimension to the fitness dimension, and proposes the use of decomposed techniques to aid in checking and diagnosing fitness. The proposed approach is based on decomposing the model in single-entry single-exit components. The resulting fragments represent subprocesses within the main process with a simple interface with the rest of the model. Fitness checking per component provides well-localized conformance information, aiding the causes behind the mismatches. Moreover, the relations between components can be explored to improve the diagnosis capabilities of the analysis, identifying areas with a high degree of mismatches, or providing a hierarchy for a zoom-in zoom-out analysis.

Finally, the book proposes two main applications of the decomposed approach. First, the theory is extended to incorporate data information for fitness checking in a decomposed manner. Second, a real-time event-based framework is presented for monitoring fitness in an on-line setting.

This book combines both application and research perspectives. It provides concrete use cases that illustrate the problems addressed by the techniques in the book, but at the same time, it contains complete conceptualization and formalization of the problem and the techniques, and through evaluations on the quality and the performance of the proposed techniques. Hence, this book is aimed at business analysts willing to improve their organization processes, and also data scientists interested in the topic of process-oriented data science.

April 2016   Jorge Munoz-Gama

Acknowledgments

The first person I would like to thank is my advisor, Josep Carmona. I would like to express my gratitude to him for his expertise, understanding, and patience. I appreciate his knowledge and skills (an "oracle" to me), his determination and patience in teaching me, his concept of ethics in the academic world, the hours of work spent together, and his comprehension in my darkest hours. From him I learned to never be satisfied with mediocrity, and to always give the best of myself. And the deeper I go in the academic world, the more I realize how outstanding Josep was as an advisor. Thanks for having your office always open for me, even today!

This book would not have been possible without Marcos Sepúlveda. He is the best partner one could have to co-direct this new academic adventure of Process Mining UC. He was one of the main reasons I took the opportunity that appeared here in Chile, and he has helped, guided, and pushed me since then, professionally and personally. Whenever I feel lost, overwhelmed, or pessimistic, he is there for me (with his acerbic humor). And the same can be said about our group. Sharing my time with them is great. I would like to thank Jonathan Lee, who made the effort to proofread this thesis. And I would like to extend my thanks to my colleagues, inside and outside the department, and friends here in Chile, especially to Mar Pérez-Sanagustín.

I would also like to thank Wil van der Aalst for opening the door of TU Eindhoven, making me feel a member of his group from the very first minute, and for sharing with me his views on the process mining field. It was (and still is) a real privilege to work with him. I would like to extend my thanks to the rest of the group in Eindhoven: Each visit there is an intellectual pleasure, and a personal dream. To all of you, thanks for making my stays in Eindhoven memorable.

During my PhD and afterward I have had the privilege of working with great people from other research groups around the world. I would like to thank Isao Echizen from the NII for advising me and giving me the opportunity to work in a place so exotic for me like Japan. His creative mind and his practical vision of security and the world in general is remarkable. I would like to also thank Seppe vanden Broucke (hard-working researcher and even greater person), Bart Baesens, and Jan Vanthienen from KU Leuven. Some colleagues visited us in Barcelona and Santiago, and I had the opportunity to work side by side with them: Joel Ribeiro, Thomas Stocker, Xixi Lu, and Andrea Burattin. It has been a pleasure to have you there and here. Collaborating with Ernest Teniente and Montse Estañol is clear evidence that in science there are more things in common between us than things keeping us apart.

The academic path is a road full of bumps. And therefore I would like to thank everyone who made this work possible: The people in charge of the department, the university, the research projects, and the administrative tasks. Not to mention Ramon Xuriguera and Daniel Alonso, who were always there when my deadlines were around the corner.

To enumerate all the friends who shared with me this period of my life would require another thesis: the friends of S108 (and S106, of course), The Netherlands, Japan, Bergen, Barcelona, Folloneres, China, Haskell, Castellón, San Francisco, Vic, Ragú, Gelida, Ibiza, Cantabria, Indonesia, Leuven, Italy, Iran, Curitiba, Chile, Costa Rica, Ukraine, Russia, Porto, France, and all the others. Without you, I would have finished my thesis much earlier.

As mentioned, this book comes as an elaborated version of my PhD thesis. I am particularly thankful to the organizers and jury of the *Best Process Mining Dissertation Award*: Marcello La Rosa, Antonella Guzzo, and especially Dirk Fahland, whose detailed comments and precise suggestions made this book possible.

Finally, I would like my last thanks to go to my parents, Loly and Emilio, my brother, Alex, and the rest of my family. I am truly grateful to have you by my side, in the good but also in the bad moments. This book is dedicated to you!

Contents

Chapter 1
Introduction

This chapter presents the concepts of processes, process models and event data, and provides an overview of the discipline that uses event data to improve process models, known as process mining. Moreover, the chapter introduces the reader to conformance checking – final goal of this book– the set of process mining techniques that focus on evaluate the difference between the assumed process model and the real process.

1.1 Processes, Models, and Data

There is a wide range of ways to define the concept of *process*. A simple but intuitive alternative is to define a process as *a sequence of activities performed in a specific order to achieve a specific goal*. And when this definition is considered one realizes that processes are everywhere in our daily lives. Processes are crucial parts of our industries and organizations. An assembly line in a car plant, or the management of packages in a delivery company are examples of processes in industrial scenarios. However, the concept of process is wider than that, being present in any daily activity: buying groceries in the city market is, by itself, a process. When you go to a market you visit the different stands in a given order. The stands where you buy may depend on your needs that day (e.g., groceries for yourself or to prepare a big meal for some guests), or even on the date itself (e.g., the butcher closes on Monday). The order of the stalls is determined by a set of factors (e.g., buy in stalls close to each others to avoid crossing the market each time, or buy heavy purchases at the end to not carry them around). Even concurrent activities may appear (e.g., send your son to buy bread while you buy the fish). Hence, buying groceries is a complete process with all its elements. Other examples of processes are the application for a credit card, boarding and security procedures before a flight, or the preoperative steps before a surgical procedure.

Most organizations, large and small, document their processes using *process models* by regulations such as ISO 9001. However, most of the organizations go

© Springer International Publishing AG 2016
J. Munoz-Gama: Conf. Check. ... in Process Mining, LNBIP 270, pp. 1–8, 2016.
DOI: 10.1007/978-3-319-49451-7_1

beyond static documents, and use the process models as a direct input to manage, coordinate, monitor, and validate all the activities of the organization. Fields such as Business Process Management (BPM) and Business Process Automation (BPA) are good examples of process techniques that rely on process models. Therefore, there is a need for updated models that reflect the *current* execution of the organization processes.

The times of paper and ink are progressively passing, and nowadays more and more organizations are supported by some kind of IT system. The support of the IT system in small organizations is usually passive, like the use of simple spreadsheets to keep the accounts of a small business. In other cases, like in banking, this support plays a more active role where each action is strictly controlled and monitored by the IT system. But most IT systems used in practice have something in common: the possibility to keep some kind of record of the actions occurred during the executions of the process they support. These records are called *event logs*.

Consider, for example, a doctoral scholarship process of some university. Doctoral students apply for a scholarship submitting their academic record, the CV of their advisor, and a description of their proposed project. Then the university evaluates the documentation and decides if they should grant the scholarship. The process starts by collecting all the documents and starting the procedure (*Start Processing*). Then, the three documents are evaluated in any order (*Evaluate Academic Record*, *Evaluate Advisor CV* and *Evaluate Project*). Once all documents have been evaluated, a final evaluation is done by the chief of the department (*Final Evaluation*) and the application is either accepted (*Accepted*) or rejected (*Rejected*). Finally, the decision taken is notified to the student (*Notify Results*). Table 1.1 shows an example of an event log recorded by the IT system supporting this scholarship process. Each row represents an *event* of the log. All events relating the application of a particular student form a *case*, and the events within a case are ordered by the time of their occurrence.

Notice that the information contained in an event log represents an unbiased reflection of the real process as it has been executed in the past. This contrasts with the possible information represented in a hand-made process model, where the perspective of the process analyst or the process owner could lead to an unfaithful representation of the process. The use of inaccurate models could result in incorrect predictions, wrong decisions, or poor performances. Therefore, *Process Mining* techniques focus on the use of unbiased information contained in the event data to support the management of process models.

Case	Event	Timestamp	Activity	Employee	Student ...
1	1	01-01-2014 10:00	Start Processing	Merce	Alex ...
1	2	01-01-2014 11:30	Evaluate Academic Record	Fernando	Alex ...
1	4	01-01-2014 13:30	Evaluate Project	Fernando	Alex ...
1	8	01-01-2014 17:00	Evaluate Advisor CV	Fernando	Alex ...
1	9	02-01-2014 10:00	Final Evaluation	Ana	Alex ...
1	11	02-01-2014 11:00	Accept	Ana	Alex ...
1	12	02-01-2014 12:00	Notify Results	Merce	Alex ...
2	3	01-01-2014 12:00	Start Processing	Merce	Dolores ...
2	5	01-01-2014 14:00	Evaluate Academic Record	Maria	Dolores ...
2	6	01-01-2014 15:00	Evaluate Advisor CV	Maria	Dolores ...
2	7	01-01-2014 16:00	Evaluate Project	Fernando	Dolores ...
2	10	02-01-2014 10:30	Final Evaluation	Ana	Dolores ...
2	13	02-01-2014 13:00	Reject	Ana	Dolores ...
2	15	02-01-2014 16:00	Notify Results	Merce	Dolores ...
3	14	02-01-2014 15:00	Start Processing	Merce	Emilio ...
3	16	02-01-2014 17:30	Evaluate Academic Record	Maria	Emilio ...
3	17	02-01-2014 18:00	Evaluate Project	Maria	Emilio ...
3	18	03-01-2014 10:00	Final Evaluation	Ana	Emilio ...
3	19	03-01-2014 11:00	Reject	Ana	Emilio ...
3	21	03-01-2014 13:00	Notify Results	Merce	Emilio ...
4	20	03-01-2014 12:00	Start Processing	Merce	George ...
...	...	...	...	...	... ...

Table 1.1: Example of a partial event log for the scholarship process, grouped by case.

1.2 Process Mining

Process mining is a relative young research discipline that sits between machine learning and data mining on the one hand, and process modeling and analysis on the other hand. The idea of process mining is to discover, monitor and improve real processes by extracting knowledge from event logs readily available in today's IT systems [7].

Event logs are unbiased *footprints* representing the process *as it is*. This contrasts with the process *assumed* by the process owners, a perception possibly biased by their understanding of how the process is executed. Event logs are the starting point of all process mining techniques, that use them to discover, verify or extend models for the process. Figure 1.1 shows an overview of how process mining is structured.

Process mining techniques can be grouped into three classes depending on their purpose: *discovery*, *conformance*, and *enhancement*.

Discovery: A discovery technique takes an event log as input and produces a model considering only the knowledge extracted from the log. For example, the model in Figure 1.2 shows a possible workflow model discovered from the scholarship process event log of Table 1.1. For simplicity purposes, the model is de-

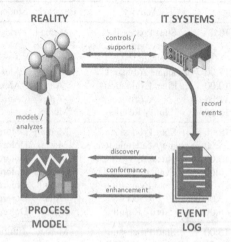

Fig. 1.1: Overview of process mining and its three types of techniques: discovery, conformance and enhancement [7].

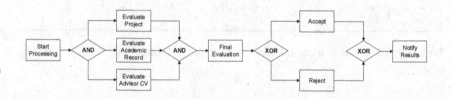

Fig. 1.2: Informal process model of the scholarship process.

fined using an informal but intuitive notation, instead of wide-spread process formalisms such as BPMN [66], Petri nets [65], or YAWL [11]. In this informal notation *AND* diamonds model concurrent activities and *XOR* diamonds represent mutually exclusive activities, similar to the gateways of BPMN. Discovery is the oldest sub-discipline in process mining. The list of approaches is long, as long as the list of different algorithms used to discover models. For instance, α-algorithm [15] is based on detecting relations among activities of the log, Genetic Miner [54] uses evolutionary algorithms, or Genet [38] and ILP Miner [93] are based on the theory of regions. The model discovered is not restricted to the workflow perspective of the process. If the log contains additional information one can also discover resource-related models, for example capturing the interactions between the actors of the process [13].

Conformance: Unlike discovery, conformance considers both an event log and a model as inputs. The goal of conformance techniques is to check whether if the modeled behavior accurately represents the behavior observed in the log. Some examples of conformance techniques are [73, 17, 59, 62, 88]. Conformance can be used to verify if a process documentation is still updated, or to check if the model obtained using a discovery approach is a faithful representation of the process.

Enhancement: Like conformance, enhancement also considers both an event log and a model as inputs. However, enhancement techniques aim to improve the a-priori model with information contained in the log. This enhancement could be done by means of *repairing* the model to better conform the observed behavior [44]. Another option is to extend the model with additional information, adding a new perspective to the process model. For example, detecting and including the data dependencies that affect the routing of a process execution [76].

This book focuses on conformance checking, and how this techniques can be used, for example, to verify the quality of hand-made models, to compare them with the unbiased execution of the processes in the event logs, and to identify possible points for improvement.

1.3 Conformance Checking Explained: The University Case

Most of the research in process mining has mainly focused on discovery techniques, neglecting the importance of conformance. However, conformance techniques have become an essential part in the process life cycle. Regulations, certifications or simply communication purposes are making most organization document their processes. Process models are also required in process-aware information systems such as workflow management systems, or for analyzing and simulating future scenarios. However, models that do not adequately represent the reality could lead to wrong conclusions.

We illustrate how conformance checking techniques can overcome such problems in the analysis of real-life processes by means of an example – *The University Case* – and we will return to this example at various points in the book:

An American university has most of their daily academic processes partially managed by its own process management system. The system was initially developed in the early 2000s, and it has been patched and extended until now. However, the cost of maintaining this system is extremely high, and not flexible at all, so the university decides to change the system that supports the academic processes.

Instead of develop a new system, the university decides to buy *Usoft*, a BPM solution for large academic institutions used in a wide range of universities in America, and configure it to fulfill their needs. Usoft software uses process models as inputs, and these models are used to guide the user step-by-step on its execution, avoiding not-desired arbitrary additional behavior. Moreover, the process models are used to automatize certain steps, and to provide monitoring functionality for the university administration.

In addition to the software, the Usoft package provides a pack of standard process models for the typical university processes to be used as input, e.g., inscription of courses, admission of new students, assign classrooms to courses, among other examples. The pack includes different process models for the same process, representing good but different procedures to perform the same process. Moreover, the process models are easily modifiable, in order to add, remove, or change certain steps.

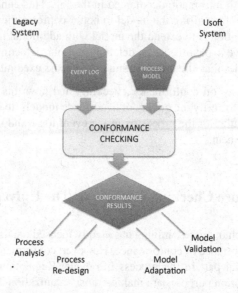

Fig. 1.3: Overview of the conformance analysis applied to the University case.

The new system should be as close as possible in its behavior as the old system. Thus, the university has to identify process variants among the provided processes which best support the process on how it has been executed in the past. For that task, the university relies on conformance checking techniques in order to successfully migrate and configure the new system, as it is shown in Figure 1.3. To do that, the university will use the event logs recorded in the legacy system to validate the models used as input in Usoft, and also to analyze how they are currently executing certain processes. In particular, some of the tasks performed with the help of conformance checking are:

- Assess the quality of the process models provided by Usoft in order to choose the one closest to the current process.
- Locate the discrepancies between a selected model and the current execution of the process, to evaluate possible modifications of the model.
- Compare the current process executions with the new process models in order to detect parts of the process not currently supported by the legacy system, and to estimate the potential benefits (time, resources, ...) of the new process automation using Usoft.

- Compare the current execution of processes with well established and widely used process models, to consider if necessary, re-design (totally or partially) the process and the university policies.

How to conduct such conformance analysis under various conditions and circumstances is the subject of this book. In Chapter 2, we introduce conformance checking at a more conceptual level and the challenges that will be addressed in this book.

1.4 Book Outline

This book is divided into four parts:

Part I - Conformance Checking in Process Mining

Part I presents the problem of conformance checking, where a process model is compared with event data to assess its conformance to the reality (Chapter 2), and the elements involved in the problem such as process models and event logs (Chapter 3)

Part II - Precision in Conformance Checking

Part II focuses on precision checking, the problem of analyzing if the process model captures the reality in a precise way, not allowing additional behavior that are not observed in the event log. Part II motivates the importance of precision checking (Chapter 4), proposes an approach to evaluate precision (Chapter 5), and shows its applicability (Chapter 6). Moreover, the precision analysis is extended to address some of the challenges faced by real-life analysis, such as handling noise and incompleteness (Chapter 7) assessing the severity of the detected problems (Chapter 8), handling non-fitness (Chapter 9 and Chapter 10), and the applicability of the analysis on non-fitting scenarios. Chapter 11).

Part III - Decomposition in Conformance Checking

Part III focuses on fitness checking – measuring how much of the observed behavior is being captured by the process model – for large processes. Part III proposes a decomposition strategy (Chapter 12) and a fitness checking based on that decomposition strategy (Chapter 13). Moreover, Part III illustrates the applicability of the decomposed analysis in practice (Chapter 14) and proposes techniques to aid the diagnosis of conformance problems (Chapter 15). Additionally, Part III extends conformance checking to data-aware process models (Chapter 16) and proposes a decomposed analysis (Chapter 17). Finally, Part III illustrates the use of decomposed conformance checking techniques on real-time scenarios (Chapter 18).

Part IV - Conclusions and Future Work

Part IV concludes the book, summarizing the contributions and proposing directions for future work on the field of conformance checking (Chapter 19).

Part I
Conformance Checking in Process Mining

Chapter 2
Conformance Checking and its Challenges

An analysis of a process is as good as the models used for such analysis. This chapter provides a basic overview on conformance checking. In particular it concentrates on the quality dimensions of a process model with respect to reality, and the challenges that arise when one tries to assess the conformance of a model to reality. This chapter closes with an overview on all the challenges addressed in this book and the respective chapters where each challenge will be covered.

2.1 The Role of Process Models in Conformance Checking

Process models play a crucial role in any process analysis technique. The term process model may refer to any representation, generic or specific, of one or several perspectives of a process. However, one of the most extended meanings of process models are workflow process models, i.e., a process model that captures the order of the actions involved in the process. Workflow process models are the main process modeling type used in this book, and they are referred simply as process models. Figure 2.1 is an example of process model, using an informal modeling notation, for the scholarship process presented in the previous chapter.

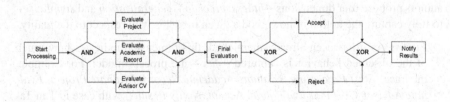

Fig. 2.1: Informal process model of the scholarship process.

© Springer International Publishing AG 2016
J. Munoz-Gama: Conf. Check. ... in Process Mining, LNBIP 270, pp. 11–18, 2016.
DOI: 10.1007/978-3-319-49451-7_2

Different process modeling notations define different types of elements to represent the process. However, there are elements common in most of the notations. These are the *activities* denoting the steps of the process, and usually graphically represented by boxes. Another element common in most process modeling notations are the *flows* between activities, represented as arrows, to denote an ordering relation between activities. Most process models are able to represent the idea of *concurrency* (several activities performed without an specific order), and *choice* (the execution of an activity excludes another). For example, in Figure 2.1 the AND and XOR gateways represent concurrency and choice, respectively. In the context of conformance checking, a process model may either be created by a human modeler or be constructed by an algorithm. Conformance checking then aims to answer how well that model describes reality - as it has been recorded in an event log.

2.2 Dimensions of Conformance Checking

By measuring the conformance between an event log and a process model one is concerned about quantifying if a given model is a valid description of reality. A first naive approach/could be to consider that a model and a log conform to each other simply if the model captures all the behavior observed in the log. In other words, a perfect conformance would require that all the traces in the log *fit* in the model. However, there are models that will allow any log to fit, but that have such a trivial structure that they are of little or no use to a process analyst when trying to understand a process. For example, let us consider the model in Figure 2.2 for the scholarship example. The informal semantics behind this model (similar to Petri nets), know as a *flower model*, is that it captures a possible sequence of the activities, in any order and for any length, i.e., the special circle in the middle should be read as the state the process is in and always return to after executing an activity. Therefore, any possible log involving the same activities fit this model. However, as one can see, this model provides absolutely no insight into the process or how the activities are executed. This simple counter-example shows that conformance needs to consider more dimensions than fitness to give a faithful account of how well a model describes a log.

In [77, 73] the multidimensional nature of the conformance is studied, and the authors propose four dimensions – *fitness*, *precision*, *generalization* and *simplicity* – to fully capture the notion of how good a given model is with respect to the reality.

Fitness As it has been already mentioned, this dimension indicates how much of the observed behavior is captured – *fits* – the process model. For example, the trace ⟨*Start Processing, Evaluate Academic Record, Evaluate Project, Evaluate Advisor CV, Final Evaluation, Accept, Notify Results*⟩ with case id 1 in Table 1.1 perfectly fits the model in Figure 1.2. However, the trace ⟨*Start Processing, Evaluate Academic Record, Evaluate Project, Final Evaluation, Reject, Notify Results*⟩ with case id 3 does not fit the model because evaluate advisor CV is never executed, denoting that the application of the student is rejected

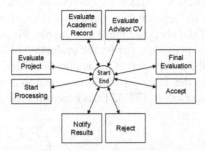

Fig. 2.2: Informal *flower* process model of the scholarship process, modeling any possible sequence of the activities.

without proper evaluation. On the other hand, both traces fit the flower model of Figure 2.2. Part III of this book is devoted to analyze the fitness dimension in a decomposed way, and consequently a more formal presentation of the fitness dimension is presented.

Precision This dimension identifies overly general models: precision penalizes a process model for allowing behavior that is unlikely given the observed behavior in the event log. For example, in the log of Table 1.1 we observe that, although the three documents could be evaluated concurrently, the university employees always first evaluate the academic record. That way, if the student is clearly not suitable for the grant (e.g., the grade does not reach the minimum necessary), the advisor and project evaluation can be done less thoroughly without compromising on the outcome of the evaluation. However, because of that specific order, the model of Figure 2.1 is less precise than reality as it also allows for other unseen execution orders. In contrast, the model shown in Figure 2.3 is a more precise representation of reality than Figure 2.1. The flower model in Figure 2.2 is the perfect example of completely imprecise model. Part II of this book is devoted to the precision dimension, and consequently a more formal presentation of the precision is included in these sections.

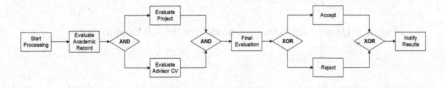

Fig. 2.3: More precise model for the scholarship process.

Generalization This dimension addresses overfitting models: a good model must be able to generalize and reproduce possible future behavior, instead of capturing simply each trace of the observed log. For example, Figure 2.4 shows a model that only captures one possible order for the evaluation of the documents that results necessarily in the acceptance of the application. This model perfectly captures the first trace in the Table 1.1, but it is unable to generalize for any other possible process execution. In [73, 77] the generalization dimension is covered in more detail.

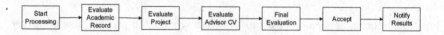

Fig. 2.4: Model that overfits the first trace of the scholarship log, and does not generalize for possible future behavior.

Simplicity This dimension penalizes models that are unnecessarily complex: following the Occam's Razor principle, models that explain the behavior observed in the log in a simple way are preferred than those that use redundant components. Figure 2.5 illustrates an example where explicitly writing out all possible execution sequences of the three evaluate activities complicates the model for the scholarship process unnecessarily. In [73, 77] the simplicity dimension is covered in more detail.

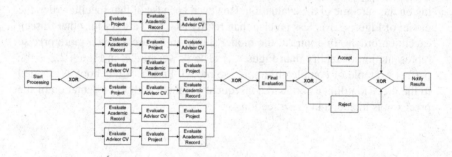

Fig. 2.5: Unnecessary complex model for the scholarship process.

Given the orthogonal nature of the dimensions, there is no such thing as perfect model, but a set of suitable models. For example, for analyzing the main paths of a organization process the analyst could prioritize fitness over the the other dimensions. On the other hand, if the process model involves critical activities and it is being used as part of a workflow system, a model with high precision is desired to avoid performing costly actions in the wrong moments.

Next, we outline the basic techniques used in conformance checking, and the challenges addressed in this book - focusing on fitness and precision.

2.3 Replay-based and Align-based Conformance Checking

In early works on conformance, most of the proposed approaches were based on *replaying* the log on the model to detect discrepancies. Some replay-based approaches simply stop at the point where the model is not able to reproduce the trace anymore. Other approaches perform the replay in a non-blocking way, regardless of whether the path of the model is followed or not, like [77]. More sophisticated approaches, such as the approach in [90], include also a *look ahead* function to determine the most promising path. Recently, another family of approaches has appeared, where the conformance check is done in a global manner, by means of *aligning* both the modeled behavior and the behavior observed in the log. Examples of conformance approaches based on alignments are [17, 50]. These approaches handle conformance in a global way, but they are computationally more expensive compared with replay-based approaches. In Part II of this book, both replay-based and alignment-based approaches are explored to check precision. In Part III, a decomposed technique is proposed to alleviate computation time for conformance diagnosis, especially for those analyses based on alignments due their expensive cost.

2.4 Challenges of Conformance Checking

Conformance checking must confront a set of challenges in order to be applied successfully. In particular, we identify five challenges: *four-dimensional conformance, big data and real time, noise, incompleteness, unfitness, and indeterminism, conformance diagnosis* and *modeling notations*. This book addresses all these challenges (see Figure 2.6).

> ***Challenge 1*** *– Four-Dimensional Conformance.* Since the multidimensional nature of conformance *– fitness, precision, generalization and simplicity –* has been stated first in [75] and later refined in [78, 77, 73], the relation between the four dimensions and the adequacy of the results has become more and more clear. Works like [35] illustrate the need of metrics for all the dimensions in order to discover good models. However, most of the approaches proposed in conformance, especially on the early days, are focused exclusively on fitness. Conformance checking must provide also measures for other dimensions such as precision, generalization, and simplicity. Hence, the challenge addressed in this book is to provide a versatile, well founded, yet easy to understand way to measure precision. This challenge is addressed in Chapters 4, 5, and 6.

Challenge 2 – Big Data and Real Time. The amount of information recorded by the information systems periodically grows exponentially. Event logs become more detailed, complete and large, and with them also the process models. Conformance techniques must evolve accordingly in order to handle this exponential growth, especially those based on the global aligning of behaviors. Moreover, the fast implantation of online and monitoring paradigms in nowadays systems is requiring faster and more fine-grained conformance approaches. In this book, we will address that challenge proposing approaches to measure conformance even on very large models and very large data sets. This challenge is addressed in Chapters 12, 13, 14 and 18.

Challenge 3 – Noise, Incompleteness, Unfitness, Indeterminism. Typically, the application of process mining techniques faces some of these four issues: noise, incompleteness, unfitness, and indeterminism. *Noise* in event logs can appear by traces incorrectly recorded (for instance, due to temporary system failure), or traces reflecting exceptional situations not representative of the typical behavior of the process. Noise is a well-known problem in discovery approaches [7], and therefore, conformance approaches proposed should also be noise-aware too. Conformance checking compares reality and model, and therefore, the comparison is only fair if the log really is *complete* regarding what happens in reality, e.g., comparing a small sample from reality to a complex models could lead to incorrect conclusions. However, assuming that a sample log contains all possible behavior is an unrealistic assumption in most of the cases. The number of traces necessary for a complete log grows exponentially when the number of concurrent actions in the model is increased. Moreover, some concurrent actions may look sequentially in the log because performing one action is always much faster than the other. Conformance techniques must include mechanisms to aid the process analyst on deciding whether the problems are real conformance anomalies or result of the incompleteness of the log. *Unfitness* – i.e., situations where the behavior observed in the log cannot be reproduced by the model – is a conformance dimension itself, but it may influence other dimensions: if the model cannot reproduce the observed behavior, it cannot determine the state of the system in that moment. Conformance approaches should try to abstract from how the alignment between observed and modeled behavior is done. This include also the *non-deterministic situations* produced when a trace in the log can be mapped to several sequences in the model. In this book, we will present conformance techniques that mitigate the effects of noise, incompleteness, unfitness, and non-determinism, providing at the same time useful conformance assessment of the process models. This challenge is addressed in Chapters 7, 9, 10, and 11.

Challenge 4 – Conformance Diagnosis. The importance of indicating the location of the problems for a proper conformance diagnosis was already emphasized in the seminal work [73]. However, the diagnosis mechanisms cannot be limited to simply locate the possible conformance errors, but they must go a step further: they must provide mechanisms to the analyst to fully understand the causes of the problems. For example, making it possible to dynamically inspect the conformance results at different levels of abstraction, or to group mismatches with a similar root cause. Diagnosis tools are especially useful for large models or models with a high degree of complexity, where the causes of the problems are difficult to grasp. In this book, we will complement the conformance techniques, with additional approaches to analyze, locate, and rank, the conformance discrepancies detected, aiding on understanding the underlying causes. This challenge is addressed in Chapters 8 and 15.

Challenge 5 – *Modeling Notations*. Most of the approaches presented in conformance so far focus exclusively on the control-flow perspective of the process – i.e., the order of the activities – and to one specific workflow modeling notation, *Petri nets* [65]. Conformance techniques must include other modeling notations, and other perspectives. In addition, there will appear new approaches to check the conformance of multi-perspective models – models capturing more than one perspective – like for example [50], where integer linear programming techniques are used to validate both the control-flow and the data perspectives of a model. In this book, we will go a step in that direction, providing conformance checking techniques adapted for data-aware multi-perspective models, especially suitable for large processes. This challenge is addressed in Chapters 16 and 17.

Figure 2.6 provides an overview of the approach presented on this book, and the techniques proposed to address each one of the challenges aforementioned.

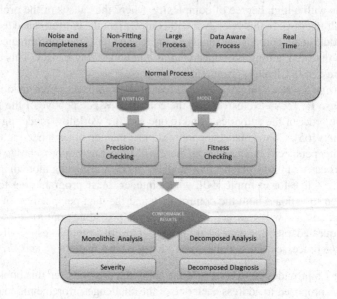

Fig. 2.6: Overview of the conformance analysis and challenges addressed in this book.

Chapter 3
Conformance Checking and its Elements

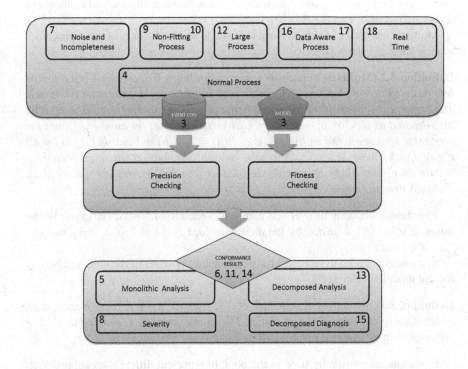

This chapter provides a basic overview on the elements involved in a conformance checking problem. In particular, the chapter introduces basic notation, and formalize the concepts of event logs, and process models. In later chapters, we will use these formalization of event logs and process models to precisely define specific problems and solutions in conformance checking.

© Springer International Publishing AG 2016
J. Munoz-Gama: Conf. Check. ... in Process Mining, LNBIP 270, pp. 19–29, 2016.
DOI: 10.1007/978-3-319-49451-7_3

3.1 Basic Notations

In this section we introduce the basic notations for sets, multisets, functions, sequences, and transition systems.

Definition 3.1 (Set) *A set A is a possible infinite collection of elements. The elements in the set are listed between braces, e.g., $A = \{a,b,c\}$. The empty set is represented by $\emptyset$. $|A|$ denotes the size of the set, e.g., $|A| = 3$. $\mathscr{P}(A)$ denotes the powerset of A, the set of all subsets of A, including the empty set and A itself, e.g., $\mathscr{P}(A) = \{\emptyset, \{a\}, \{b\}, \{c\}, \{a,b\}, \{a,c\}, \{b,c\}, \{a,b,c\}\}$.*

Some operations are defined over sets. Let $A = \{a,b,c,d\}$ and $B = \{a,c,d,e\}$ be non-empty sets. The *union* of A and B, denoted $A \cup B$, is the set containing all elements of either A or B, e.g., $A \cup B = \{a,b,c,d,e\}$. The *intersection* of A and B, denoted $A \cap B$, is the set containing elements in both A and B, e.g., $A \cap B = \{a,c,d\}$. The *difference* between A and B, denoted $A \setminus B$, is the set containing all elements of A that are not in B, e.g., $A \setminus B = \{b\}$.

Multisets are used to describe event logs and to represent states of some processes models such as Petri nets.

Definition 3.2 (Multiset) *A multiset – also called* bag *– B over a set A is a possible infinite collection of elements of A, where each element may appear more than once. The elements in the multiset are listed between square brackets, e.g., $B = [a,a,b]$, also denoted as $B = [a^2, b]$, where $A = \{a,b\}$. $B(a)$ denotes the number of times the element a appears in the multiset B, e.g., $B(a) = 2$, $B(b) = 1$ and $B(c) = 0$ for all $c \notin A$. Furthermore, a set $S \subseteq A$ can be viewed as a multiset where each element occurs once. The empty multiset is denoted as $[\,]$. $\mathscr{B}(A)$ represents the set of all multisets over the set A.*

The standard operations over sets can be extended to multisets. For example: the *union* $[a,a] \uplus [a,b] = [a,a,a,b]$, the *difference* $[a,a,b,c] \setminus [a,b,b] = [a,c]$, the *size* $|[a,a]| = 2$, etc.

Sequences are used to represent traces in the log, or states of the observed behavior and modeled behavior.

Definition 3.3 (Sequence) *A sequence $\sigma = \langle a_1, a_2, \ldots, a_n \rangle \in A^*$ over a set A is an ordered list of elements $a_i \in A$ of size n. The empty sequence is denoted as $\langle \rangle$. The concatenation of sequences is denoted as $\sigma_1 \cdot \sigma_2$.*

Functions are constantly used in the book to represent different meanings, such as the mapping between model tasks and observable activities.

Definition 3.4 (Function) *Let A and B be non-empty sets. A function f from A to B, denoted $f : A \to B$ is a relation from A to B, where every element of A is associated to an element of B. Given a function $f : A \to B$, $dom(f)$ and $rng(f)$ denote the domain and the range of f. A partial function, denoted $f : A \nrightarrow B$, is a function that may be undefined for some of the elements in A.*

In particular, a function can be applied to a sequence, applying it only to the elements in the domain of the function, i.e. if $dom(f) = \{x,y\}$, then $f(\langle y,z,y\rangle) = \langle f(y), f(y)\rangle$.

Definition 3.5 (Function to Sequences) *Let $f \in X \nrightarrow Y$ be a partial function. f can be applied to sequences of X using the following recursive definition (1) $f(\langle\,\rangle) = \langle\,\rangle$ and (2) for $\sigma \in X^*$ and $x \in X$:*

$$f(\langle x\rangle \cdot \sigma) = \begin{cases} f(\sigma) & \text{if } x \notin dom(f) \\ \langle f(x)\rangle \cdot f(\sigma) & \text{if } x \in dom(f) \end{cases}$$

An especially useful function is the *projection*.

Definition 3.6 (Projection) *Let X be a set and $Q \subseteq X$ one of its subsets. $\lceil_Q \in X^* \to Q^*$ is a projection function and is defined recursively: (1) $\langle\,\rangle\lceil_Q = \langle\,\rangle$ and (2) for $\sigma \in X^*$ and $x \in X$:*

$$(\langle x\rangle \cdot \sigma)\lceil_Q = \begin{cases} \sigma\lceil_Q & \text{if } x \notin Q \\ \langle x\rangle \cdot \sigma\lceil_Q & \text{if } x \in Q \end{cases}$$

For example, $\langle y,z,y\rangle\lceil_{\{x,y\}} = \langle y,y\rangle$.

Transition systems are used to represent the observed and the modeled behavior. Moreover, in order to abstract from any specific process model notation, transition systems are used to represent the semantics of a generic process model.

Definition 3.7 (Transition System) *A transition system with initial states S_0 and final states S_F is a tuple $TS = \{S, \Sigma, \nearrow, S_I, S_F\}$ where S is a set of states, Σ is a set of action labels, $S_I \subseteq S$ is a set of initial states, $S_F \subseteq S$ is a set of final states, and $\nearrow \subseteq S \times \Sigma \times S$ is a transition relation.*

A more general – and more commonly used – definition of transition systems does not include the set of final states. When no set of final states is specified, it is assumed all states are final states, i.e., $S_F = S$. Figure 3.1 shows an example of transition system with states $S = \{s_1, s_2, s_3, s_4\}$ and labels $\Sigma = \{a,b,c\}$, being s_1 its initial state. The transition system denotes that, for example, there is a transition from the state s_1 to the state s_2.

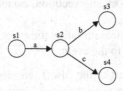

Fig. 3.1: Example of TS with $S = \{s_1, s_2, s_3, s_4\}$, $\Sigma = \{a,b,c\}$, and $S_I = \{s_1\}$.

3.2 Event Logs

Event	Case	Timestamp	Activity	Employee	Student ...
1	1	01-01-2014 10:00	(a) Set Checkpoint	Merce	Alex ...
2	1	01-01-2014 11:30	(c) Evaluate Academic Record	Fernando	Alex ...
3	2	01-01-2014 12:00	(a) Set Checkpoint	Merce	Dolores ...
4	1	01-01-2014 13:30	(b) Evaluate Project	Fernando	Alex ...
5	2	01-01-2014 14:00	(c) Evaluate Academic Record	Maria	Dolores ...
6	2	01-01-2014 15:00	(d) Evaluate Advisor CV	Maria	Dolores ...
7	2	01-01-2014 16:00	(b) Evaluate Project	Fernando	Dolores ...
8	1	01-01-2014 17:00	(d) Evaluate Advisor CV	Fernando	Alex ...
10	1	02-01-2014 11:00	(e) Accept	Ana	Alex ...
11	1	02-01-2014 12:00	(a) Set Checkpoint	Merce	Alex ...
12	2	02-01-2014 13:00	(f) Reject	Ana	Dolores ...
13	2	02-01-2014 16:00	(a) Set Checkpoint	Merce	Dolores ...

Table 3.1: Example of event log for the scholarship process variant, ordered chronologically.

Event logs are the footprints left in the system by the execution of processes. They are the main objects that any process mining technique works with. Let us consider a variant of the scholarship process presented in Chapter 1. In this variant both *Start Processing* and *Notify Results* activities are substituted by a single activity *Set Checkpoint* executed at the start and end of the process. Moreover, the *Final Evaluation* action is always done outside the system, and there is no recording of it in the log. Table 3.1 shows a possible partial event log of the process. The log contains two complete process executions or *cases*: Alex and Dolores application. Each row corresponds with one *event* in the log, and the events are ordered chronologically. Therefore, an event log is a sequence of events. Each event is associated with a set of attributes. The list of the most common attributes in event logs for the practice of process mining analysis are:

- *case* – process instance id of the event.
- *activity* – name of the action performed in the event.
- *timestamp* – moment of the event execution, establishing an order between the events.
- *resource* – name of the resource initiating the event.
- *data* – data attribute related to the event.

For example, the event 10 is part of the case 1, corresponding with the scholarship application of *Alex*, and it reflects an *Accept* activity performed by *Ana* at $11:00$ on the date $02 - 01 - 2014$. For the sake of clarity, activities are associated with a lowercase letter, e.g., $e = Accept$.

Different attributes are required to derive different type of models. For example, the resource attribute is necessary for discovering social interactions between actors of the process. In this book we focus mainly on the control-flow perspective of the processes (except on Chapter 17 were we also focus on the data perspective). For those cases, the activity, the case and an order between events are necessary, and therefore, the definition of event logs is simplified: the event log is composed by *traces*, where each trace corresponds to a case, and only the activity is considered for each event.

Definition 3.8 (Trace, Event Log) *Let $A \in \mathcal{U}_A$ be a set of activities in some universe of activities $\mathcal{U}_A$. A trace $\sigma \in A^*$ is a sequence of activities. An event log is a multiset of traces, i.e., $L \in \mathcal{B}(A^*)$.*

For example, the event log in Table 3.1 is represented as $[\langle a,c,b,d,e,a \rangle, \langle a,c,d,b, f,a \rangle]$, containing information about 12 events and 2 cases, where each case follows a different trace.

Although they are called event logs, event information is rarely recorded in logs files as *Apache logs* or *error logs* are, but stored in some internal database. However, recently a new XML-based standard for event logs has been presented: *eXtensive Event Stream (XES)* [1]. The purpose of XES is not only the storage of event logs, but to provide a standard format for the interchange of event log data between tools and application domains.

3.3 Process Models

Process models are the second element necessary in any conformance checking approach. A process model captures the behavior to compare with respect to the behavior observed in the log. Different models are used to capture different perspectives of the process. In this book we mainly focus on the control-flow perspective of processes, and therefore the models are used to capture the ordering between the actions. For the sake of generality, whenever it is possible, we abstract from any specific process modeling notation by defining a generic *process model*. In Section 3.4 we present several concrete process modeling languages.

A generic process model semantics can be abstracted using a basic transition systems as a placeholder for more advanced modeling languages, such as Petri nets, UML, BPMN, EPCs, etc.

Definition 3.9 (Process Model) *A process model semantic is abstracted as a transition system $TS = \{S, T, \nearrow, S_I, S_F\}$ over a set of model tasks T with states S, initial states $S_I \subseteq S$, final states $S_F \subseteq S$, and transitions $\nearrow \subseteq S \times T \times S$. When no set of final states is specified, all states are assumed to be final states.*

A task $t \in T$ is *enabled* to be executed in the state s of the model M, denoted as $(M,s)[t\rangle$, if there is a transition with task t in the state s, i.e., $(s,t,s') \in \nearrow$. $(M,s)[t\rangle(M,s')$ denotes that t is enabled in s and executing t results in the state s'. Let $\sigma = \langle t_1, t_2, \ldots, t_n \rangle \in T^*$ be a sequence of model tasks. $(M,s)[\sigma\rangle(M,s')$ denotes that there is a set of states $s_0, s_1, \ldots, s_n$ such that $s_0 = s$, $s_n = s'$ and $(M, s_i)[t_{i+1}\rangle(M, s_{i+1})$ for $0 \leq i < n$.

Definition 3.10 (Complete Task Sequences) *The set of complete tasks sequences ϕ_t of process model M is the set of all possible sequences of tasks executed from an initial state reaching a final state, i.e., $\phi_t(M) = [\sigma|(M, s_I)[\sigma\rangle(M, s_F)]$ for all $s_I \in S_I$ and $s_F \in S_F$.*

Figure 3.2 shows a process model M capturing the order of the tasks $T = t_1 \ldots t_8$ using the informal semantics of Chapter 1, i.e., *AND* models concurrent tasks and *XOR* mutually exclusive tasks. The set of complete tasks sequences of this model is:

$$\phi_t(M) = \begin{array}{l} \langle t_1, t_2, t_3, t_4, t_5, t_6, t_8 \rangle \\ \langle t_1, t_2, t_3, t_4, t_5, t_7, t_8 \rangle \\ \langle t_1, t_2, t_4, t_3, t_5, t_6, t_8 \rangle \\ \langle t_1, t_2, t_4, t_3, t_5, t_7, t_8 \rangle \\ \langle t_1, t_3, t_2, t_4, t_5, t_6, t_8 \rangle \\ \langle t_1, t_3, t_2, t_4, t_5, t_7, t_8 \rangle \\ \langle t_1, t_3, t_4, t_2, t_5, t_6, t_8 \rangle \\ \langle t_1, t_3, t_4, t_2, t_5, t_7, t_8 \rangle \\ \langle t_1, t_4, t_2, t_3, t_5, t_6, t_8 \rangle \\ \langle t_1, t_4, t_2, t_3, t_5, t_7, t_8 \rangle \\ \langle t_1, t_4, t_3, t_2, t_5, t_6, t_8 \rangle \\ \langle t_1, t_4, t_3, t_2, t_5, t_7, t_8 \rangle \end{array}$$

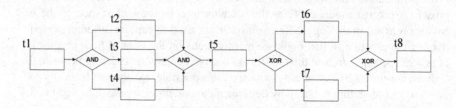

Fig. 3.2: Informal process model for tasks $t_1 \ldots t_8$.

In a *labeled process model* (or observable process model) the tasks represent activities in a real process that are potentially observable, i.e., they may cause events in a log. This potential observability is captured by the *labeling function*, which relates observable tasks in the model to activities of the process.

Definition 3.11 (Labeled Process Model) *A* labeled process model *(or simply a process model) is a tuple $M = (TS, l)$, where the transition system $TS = \{S, T, \nearrow, S_I, S_F\}$ represents the semantics of the model, and $l \in T \nrightarrow \mathcal{U}_A$ is a labeling function where $\mathcal{U}_A$ is some universe of activity labels.*

The process models in this book are considered always labeled, unless stated otherwise. If a model task $t \notin dom(l)$, then it is called *invisible* (or also known as *silent* or *unobservable*). An occurrence of a *visible* task $t \in dom(l)$ corresponds to observable activity $l(t)$. Two or more tasks associated with the same activity are called *duplicate*. For instance, Figure 3.3 show a labeled process model for the scholarship variant process. The task t_5 is invisible, i.e., this action is not observed nor recorded in the log. Tasks t_1 and t_8 are duplicate tasks corresponding with the same activity *(a)Set Checkpoint*.

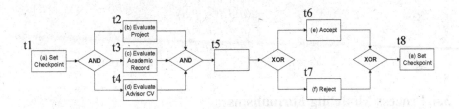

Fig. 3.3: Informal labeled process model for scholarship variant process.

Definition 3.12 (Process Model notations) *Let* $M = (TS, l)$ *be a process model with* $TS = \{S, T, \nearrow, S_I, S_F\}$.

- $T_v(M) = dom(l)$ *is the set of* visible tasks *in M.*
- $A_v(M) = rng(l)$ *is the set of corresponding* observable activities *in M.*
- $T_v^u(M) = \{t \in T_v(M) \mid \forall_{t' \in T_v(M)} \, l(t) = l(t') \Rightarrow t = t'\}$ *is the set of* unique visible tasks *in M, i.e., there are no other tasks with the same associated to the same activity.*
- $A_v^u(M) = \{l(t) \mid t \in T_v^u(M)\}$ *is the set of corresponding* unique observable activities *in model M.*

In the example model M of Figure 3.3 the set of visible tasks is $T_v(M) = \{t_1, t_2, t_3, t_4, t_6, t_7, t_8\}$ (all except t_5), and the set of unique visible tasks is $T_v^u(M) = \{t_2, t_3, t_4, t_6, t_7\}$. The set of observable activities is $A_v(M) = \{a, b, c, d, e, f\}$, while the set of unique observable activities is only $A_v^u(M) = \{b, c, d, e, f\}$.

Similar to the set of complete task sequences, a labeled process model contains its corresponding set of complete activity sequences, i.e., sequences of tasks starting from an initial to a final state projected onto the set of observable activities.

Definition 3.13 (Complete Activity Sequences) *Let M be a process model with T tasks and labeling function l. A sequence* $\sigma_v = \langle a_1, a_2, \ldots, a_n \rangle \in \mathcal{U}_A *$ *is a complete activity sequence of the model M, denoted as* $(M, s_I)[\sigma_v \triangleright (M, s_F)$ *if and only if there is a complete tasks sequence* $\sigma \in T^*$ *in M such that* $(M, s_I)[\sigma\rangle(M, s_F)$ *and* $l(\sigma) = \sigma_v$. *The set of complete activity sequences of the model M is denoted* $\phi(M)$.

The set of complete activity sequences of the model in Figure 3.3 is:

$$\phi(M) = \begin{array}{l} \langle a,b,c,d,e,a \rangle \\ \langle a,b,c,d,f,a \rangle \\ \langle a,b,d,c,e,a \rangle \\ \langle a,b,d,c,f,a \rangle \\ \langle a,c,b,d,e,a \rangle \\ \langle a,c,b,d,f,a \rangle \\ \langle a,c,d,b,e,a \rangle \\ \langle a,c,d,b,f,a \rangle \\ \langle a,d,b,c,e,a \rangle \\ \langle a,d,b,c,f,a \rangle \\ \langle a,d,c,b,e,a \rangle \\ \langle a,d,c,b,f,a \rangle \end{array}$$

3.4 Process Modeling Formalisms

There are a wide variety of process modeling formalism that match the generic process model definition of previous section. In this section we present some of these formalisms. In particular we focus on Petri nets and its extensions, the formal notation used to illustrate the process models examples of this book.

3.4.1 Petri Nets

Petri nets [65] are one of most frequently used process modeling notations in process mining. Its formal semantics, its mathematical foundation, and its inherent capacity to model concurrency in a succinct way, make Petri nets perfect to model the control-flow perspective of processes. In addition, Petri nets are supported by an intuitive graphical notation, and there exists a wide range of tools and libraries to operate with them.

Definition 3.14 (Petri Net) *A* Petri net *is a tuple* $PN = (P,T,F)$ *with* P *the set of* places, T *the set of* transitions, *where* $P \cap T = \emptyset$, *and* $F \subseteq (P \times T) \cup (T \times P)$ *the flow relation. For a node* n *(place or transition) of a Petri net,* $\bullet n$ $(n \bullet)$ *is the predecessor (successor) set of* n *in* A, *i.e.,* $\bullet n = \{n' | (n',n) \in F\}$ *and* $n \bullet = \{n' | (n,n') \in F\}$.

The set of transitions T represent the set of tasks of generic process modeling definition. An example of Petri net is shown in Figure 3.4. The transitions are represented as square nodes, while the places are represented as circles.

The states of a Petri net are called markings. The formal semantics of Petri nets are defined by the firing rule, that states the effects of firing an enabled transition.

Definition 3.15 (Petri Net Semantics) *Let* $PN = (P,T,F)$ *be a Petri net. A* marking M *is a multiset of places, i.e.,* $M \in \mathcal{B}(P)$. *A transition* $t \in T$ *is* enabled *in a*

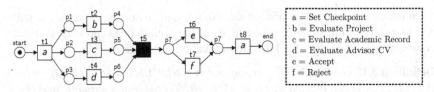

Fig. 3.4: Petri net for scholarship variant process of Figure 3.3.

marking M, denoted as $(PN,M)[t\rangle$, *iff* $\bullet t \leq M$. *Firing transition t in M, denoted as* $(PN,M)[t\rangle(PN,M')$, *results in a new marking* $M' = M - \bullet t + t\bullet$, *i.e., tokens are removed from* $\bullet t$ *and added to* $t\bullet$.

A marking is graphically represented as black dots (called *tokens*) in places. For example, the marking represented in Figure 3.4 is $M = [start]$. In that marking, only t_1 is enabled, and firing t_1, $(PN,M)[t_1\rangle(PN,M')$, will result in the marking $M' = [p_1, p_2, p_3]$. In M', t_2, t_3, t_4 are enabled simultaneously, and can be fired in any order.

Similar to tasks sequences, a transition sequence from a Petri net can also be defined. Notice that, for the sake of clarity, the same notation is preserved between generic process models and Petri nets, referring to tasks or transitions in each particular case.

Definition 3.16 (Transition Sequence) *A transition sequence* $\sigma = \langle t_1, t_2, \ldots, t_n \rangle \in T^*$ *of Petri net PN, represented as* $(PN,M)[\sigma\rangle(PN,M')$, *denotes that there is a set of markings* $M_0, M_1, \ldots, M_n$ *such that* $M_0 = M$, $M_n = M'$ *and* $(N,M_i)[t_{i+1}\rangle(N,M_{i+1})$ *for* $0 \leq i < n$. *A marking* M' *is reachable from* M *if there exists a* σ *such that* $(PN,M)[\sigma\rangle(PN,M')$.

Similar to generic labeled process models, a Petri net can also be labeled, associating observable activities to the transitions of the model.

Definition 3.17 (Labeled Petri Net) *A labeled Petri net* $PN = (P,T,F,l)$ *is a Petri net* (P,T,F) *with labeling function* $l \in T \nrightarrow \mathcal{U}_A$, *where* $\mathcal{U}_A$ *is some universe of activity labels.*

Figure 3.4 shows a labeled Petri net for the scholarship variant process of Figure 3.3. Similar to the generic labeled process model, we can define the *visible transitions* $T_v(PN)$, *observable activities* $A_v(PN)$, *unique visible transitions* $T_v^u(PN)$, and *unique observable activities* $A_v^u(PN)$ of a Petri net PN. Typically, invisible transitions are represented as filled squares, e.g., t_5.

Definition 3.18 (Activity Sequence) *A sequence* $\sigma_v = \langle a_1, a_2, \ldots, a_n \rangle \in \mathcal{U}_A*$ *is a activity sequence of the Petri net PN, denoted as* $(PN,M)[\sigma_v \rhd (PN,M')$ *if and only if there is a transition sequence* $\sigma \in T^*$ *in PN such that* $(PN,M)[\sigma\rangle(PN,M')$ *and* $l(\sigma) = \sigma_v$.

In the context of process mining and business processes, processes are usually considered to start in an initial state and to end in a well-defined end state. Petri nets considering a initial and a final marking are called *system nets*.

Definition 3.19 (System Net) *A system net is a triplet* $SN = (PN, M_I, M_F)$ *where* $PN = (P, T, F, l)$ *is a labeled Petri net,* $M_I \in \mathscr{B}(P)$ *is the initial marking, and* $M_F \in \mathscr{B}(P)$ *is the final marking.*

A system net $SN = (PN, [start], [end])$ is a possible system net for the Petri net PN in Figure 3.4. We define the set of *complete transition sequences* and the set of *complete activity sequences* as the sequence of transitions and activities from the initial marking to the final marking.

Definition 3.20 (Complete Transition Sequence, Complete Activity Sequence) *Let* $SN = (PN, M_I, M_F)$ *be a system net with* $PN = (P, T, F, l)$. *The set of complete transition sequences* ϕ_t *of SN is the set of all possible sequences of transitions executed from the initial marking and reaching the final marking, i.e.,* $\phi_t(SN) = [\sigma | (PN, M_I)[\sigma\rangle(PN, M_F)]$. *The set of complete activity sequences of system net SN is the set of all possible sequences of observable activities from the initial marking and reaching the final marking, i.e.,* $\phi(SN) = [\sigma | (PN, M_I)[\sigma \triangleright (PN, M_F)]$.

3.4.2 Workflow Nets

Workflow nets [4, 3], commonly used for business and workflow processes, are a subclass of Petri nets with a well defined starting and ending place.

Definition 3.21 (Workflow Net) *A* workflow net $WN = (P, T, F, l, start, end)$ *is a particular type of Petri net where:*

- *start is a special source place with no incoming arcs, i.e.,* $\bullet start = \emptyset$
- *end is a special sink place, with no outgoing arcs, i.e.,* $end\bullet = \emptyset$
- *every node of the net must be on some path from start to end*

The Petri net in Figure 3.4 shown in the previous section is actually a workflow net. Workflow nets present a direct way to define system nets, with a single token in *start* as initial marking, and a single token in *end* as a final marking.

Definition 3.22 (System Net from Workflow net) *A system net* $SN = (WN, [start], [end])$ *from the workflow net WN is the net where the initial marking is* $[start]$, *and the final marking is* $[end]$.

3.4.3 Other Formalisms

There is a wide range of other model formalisms to describe processes. Each formalism has its own advantages and disadvantages, and its own tools to support it. Some examples of such formalisms are *BPMN* [66], *YAWL* [11], *EPC* [5], or *UML* [67], to enumerate some examples.

Business Process Model and Notation (BPMN) [66] is a standard for business process modeling that provides a graphical notation for specifying business processes based on a flowcharting technique very similar to activity diagrams from Unified Modeling Language (UML) [67]. The objective of BPMN is to support business process management, for both technical users and business users, by providing a notation that is intuitive to business users, yet able to represent complex process semantics. BPMN is one of the most used notations in the industry. BPMN is composed of events (denoted as circles), activities (denoted as rounded squares) and gateways (denoted as diamonds), among other elements, and the connections between them. Figure 3.5 illustrates a model for the scholarship variant process using BPMN notation[1].

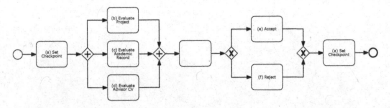

Fig. 3.5: BPMN for scholarship variant process of Figure 3.3.

Another notation is YAWL [11]. The original drivers behind YAWL were to define a workflow language that would support all (or most) of the typical workflow patterns [12] and would have a formal semantics. The language is supported by a software system that includes an execution engine, a graphical editor and a worklist handler. The system is available as Open source software under the LGPL license. Figure 3.6 illustrates a model for the scholarship variant process using YAWL notation, where the atomic tasks (denoted as squares) can be possibly complemented with control flow elements such as AND-split, AND-join, XOR-split, XOR-join, OR-split or OR-join.

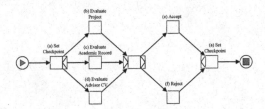

Fig. 3.6: YAWL for scholarship variant process of Figure 3.3.

---

[1] For the sake of clarity, BPMN notation is abused representing an activity without label.

Part II
Precision in Conformance Checking

Chapter 4
Precision in Conformance Checking

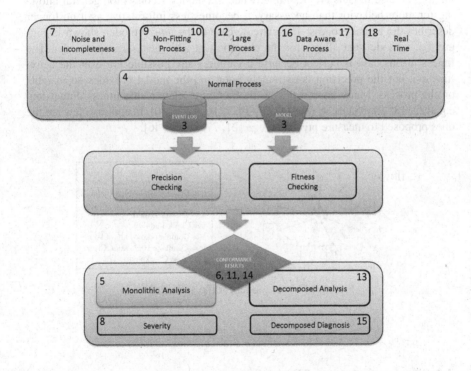

A good model must represent the reality in a precise way. This chapter provides an overview on the precision checking. In particular it concentrates on its importance as a conformance checking dimension, and it presents the requirements necessary for a precision metric. Chapter 5 will then present a solution to the problem that is validated and extended in the remaining chapters of Part II.

© Springer International Publishing AG 2016
J. Munoz-Gama: Conf. Check. ... in Process Mining, LNBIP 270, pp. 33–38, 2016.
DOI: 10.1007/978-3-319-49451-7_4

4.1 Precision: The Forgotten Dimension

In the last years, the use within organizations of Process-Aware Information Systems (PAIS) and Business Process Management technologies (BPM) has experimented an exponential growth. Increasingly, organizations are investing vast amounts of time and resources on modeling their business processes [42, 94]. Process models are used to provide insights, analyze, simulate and improve the processes, and all conclusions obtained rely on how faithfully these models describe the reality [7]. Conformance checking techniques compare recorded process executions in terms of event logs with process models to quantify how *good* these models are. Checking conformance is a complex multi-dimensional task that involves the dimensions of fitness, precision, generalization and simplicity [7]. While fitness evaluates whether the behavior in the log is captured in the model, precision evaluates how much behavior is allowed by the model which is unlikely given the observed behavior in the log. Precise models are desirable: when the model becomes too general allowing for more behavior than necessary, it becomes less informative as it no longer describes the actual process. The extreme case is the *flower* model, shown in Figure 4.1, a model that allows for the execution of activities *a–i* in any order. The fitness in this case is perfect because it captures any possible log over the activities *a–i*, but the precision is extremely poor and the model provides no insights on the process. Many approaches in the literature relate to the fitness dimension, e.g., [22, 18, 23, 35, 56, 45, 46, 68, 77, 79, 7, 10, 92, 9, 50]. In contrast, few are the ones proposed to measure precision, e.g. [21, 77, 88, 55, 46].

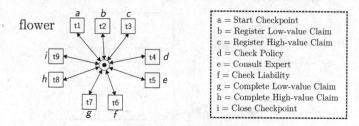

Fig. 4.1: Flower model for the activities *a–i*.

4.2 The Importance of Precision

From a theoretical point of view, the need for a multi-dimensional conformance, and especially the necessity of measuring precision, has been plenty justified and illustrated in the literature [77, 35]. The *flower model* represents the theoretical worst scenario concerning precision.

But besides the theory, the importance of precision is also a key element of real-life scenarios. The lack of precision on the models used in real-life systems may reduce the effectiveness of those systems. Some examples of those scenarios are:

- *Workflow Management Systems*: Workflow Management Systems and other process-aware information systems [42] make use of process models to set-up and monitor the defined sequence of tasks in a process. A precise model would guide the system, limiting and suggesting the next task to be performed, improving the efficiency. On the other hand, an imprecise model would be a poor guide, allowing too much tasks at the same time, and not giving real useful information.
- *Regulations and Certifications*: Regulations, such as the Sarbanes-Oxley (SOX) Act [80], enforce the documentation of processes, while quality certification, such as the ISO 9000 standards, requires the documentation and monitoring of all key processes to ensure their effectiveness. The use of overgeneralized models as part of the audition and certification procedure may provide an inexact vision of the processes, failing the auditing process [16].
- *Communication*: Models are used to communicate and gain insight into the analyzed processes, for example, models illustrating the reworks done in the process. Models allowing for more behavior than the one seen in reality would complicate the understanding of the processes, indicating a possible rework path that never happened.
- *Simulation*: Process models are used for simulating possible future scenarios, and to take decisions according to results obtained. The use of imprecise models would result in overgeneralized simulations, with a lot of non realistic scenarios, limiting the effectiveness of the conclusions.
- *Abstraction*: Some systems require a high level of flexibility. Health-care systems are good examples of flexible systems, where the path followed by two patients is never the same. In those cases, the need for precision is less crucial. However, extremely overgeneralized models would mask possible path restrictions, compromising the usefulness of the system.

In conclusion, the need for achieving precise models is becoming more crucial in nowadays systems, for both conformance checking and process discovery techniques.

4.3 Measures of Precision

In contrast with the fitness dimension, there are few proposed approaches in the literature that address, totally or partially, the precision checking. In [46], Greco et al. propose a metric –*soundness*– to estimate the precision by calculating the percentage of traces in the log that can be generated by the given model. Medeiros et al. [55] defines a metric –*behavioral precision*– to measure the precision between two models and a log, evaluating how much of the first model behavior is covered by the second. This measure is used within the *Genetic Miner* [56, 54] –a discovery approach based on evolutionary algorithms– to evaluate the quality of the population obtained. Goedertier et al. [45] introduces the use of artificial negative examples to measure the precision between an event log and a process model. The way of generating those negative examples was later improved by De Weerdt et al. [88] and Vanden Broucke et al. [30]. Finally, Van Dongen et al. [41, 40] addresses the precision dimension between two models without a log, based on the similarity of their structures.

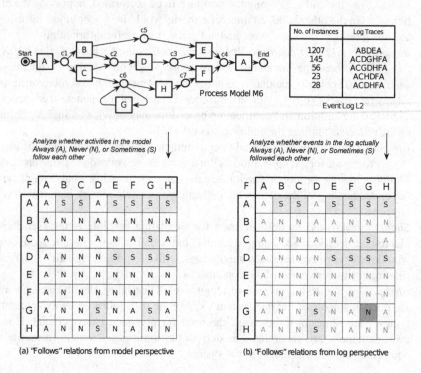

Fig. 4.2: 'Follows' relations used to compute the behavioral appropriateness (a'_B). Image taken from *Process Mining: Conformance and Extension* by Anne Rozinat [73].

However, given the goal of checking precision between a process model and an event log, Rozinat et al. [77] can be seen as the seminal work, later extended in [73]. In [77], Rozinat et al. present several metrics to estimate the four dimensions of conformance checking. In particular, they present the *advanced behavioral appropriateness (a'_B)*, a metric designed to measure the precision between a Petri net and an event log. The metric is based on deriving global 'Follows' and 'Precedes' activity relations from both a model and a log perspective, i.e., given all pairs of activities x and y, determine whether they either *always* (i.e., in all traces), *never* (i.e., in none of the traces), or *sometimes* (i.e., in some but not all traces) follow/precede each other. Figure 4.2 shows an example of 'Follows' relation, taken from [73]. Because precision relates to those situations where the model has more variability than the event log, the idea of the metric is compare the behavior allowed by the model and the behavior observed in the log based on how many elements are contained in the *sometimes follows* and *sometimes precedes* relations once we superpose two the matrices.

The aforementioned technique has some limitations. First, precision is based on activity relations with a high level of abstraction but not precise points of deviation, i.e., only three categories are considered: always, never and sometimes. Although this can be useful to get insights on the precision from a high level, it becomes a limitation when it comes to detect exactly those precision problems. Moreover, the correct categorization of follows/precedes relations is not guaranteed when the traces contain loops [73]. Finally, building the relations from a model point of view requires the analysis of the model task sequences based on a state space analysis or an exhaustive model simulation. This limits the applicability of the approach to examples of low complexity, or forces the use of approximations to alleviate the complexity.

4.4 Requirements for Precision

Given the aforementioned limitations, there are several requirements a precision technique must fulfill in order to be applied in general real scenarios, enable at the same time an analyst to both *measure* the precision dimension and *diagnose* the causes of the precision problems and their locations. In particular, a precision technique should aim for:

- *A precision based on potential points of improvement.* The precision dimension can be viewed from different angles. One can estimate the precision of a system as the difference between the behavior allowed by the model and the behavior observed in the log. Instead, we propose a different way of estimating precision based on identifying all precision problems and quantifying the effort need to correct them achieving a perfectly precise system.
- *A technique that does not require an exhaustive model state-space exploration.* Because the state space of a model can grow exponentially, state-based analysis techniques may be problematic with respect to computational complexity [72]. Therefore, any approach involving an exhaustive exploration of the model state-space, such as [77], sees its applicability compromised for complex cases, often found in reality.
- *An approach to identify precision problems with a fine granularity.* In order to help in the process diagnosis, the approach should be able to detect the exact precision points of improvement in the model.
- *A mechanism to use the precision results for analysis and process improvement.* Besides a metric, any approach proposed should consider the possibility of collecting all precision anomalies detected in a format suitable for analysis using

process mining techniques, or to be used within the continuous cycle of improvement and correction of the process.

In the following chapters, a set of precision checking approaches are presented that fulfill the aforementioned requirements.

Chapter 5
Measuring Precision

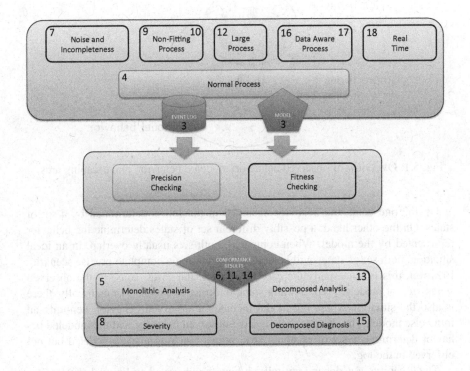

The previous chapter motivated the need for a precision measure and enumer-ated a list of requirements that the measure should satisfy. This chapter presents the procedure to measure precision, both an intuitive explanation as well as a com-plete formalization based on the definitions of Chapter 3. The procedure is based on detecting and pinpointing escaping arcs, i.e., the parts where the model allows more behavior than the one observed in the log. This basic procedure analyses the

© Springer International Publishing AG 2016
J. Munoz-Gama: Conf. Check. ... in Process Mining, LNBIP 270, pp. 39–53, 2016.
DOI: 10.1007/978-3-319-49451-7_5

model as a whole (i.e., monolithic analysis). Later chapters consider variants and extensions of the procedure regarding both its input and its output.

5.1 Precision based on Escaping Arcs

In this chapter we present a precision checking approach that fulfills all the requirements presented in Chapter 4. The approach is based on the detection and collection of escaping arcs between a log and a model. The *escaping arcs* define those crucial points where the model starts to deviate from the behavior observed in the log. Figure 5.1 shows an overview of the escaping arcs.

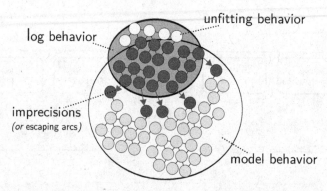

Fig. 5.1: Overview of observed behavior, modeled behavior, and escaping arcs.

On the one hand, the behavior observed in the log is determined by a set of states. On the other hand, a possibly different set of states determine the behavior represented by the model. When compared, both sets usually overlap. In an ideal situation, both set of states will be equal, representing a complete precise scenario. However, this is not usually the case. In most of the cases, some of the observed states are not modeled, representing some unfitting behavior. Symmetrically, there is also the situation where the model includes unobserved behavior, denoting an imprecise model. The 'border' between the observed behavior and the modeled behavior determines a set of escaping arcs, points reaching a state modeled but not observed in the log.

The escaping arcs depend entirely on how the observed and modeled states are determined. The technique proposed in this chapter is decomposed into the following steps (cf. Figure 5.2):

1. *Constructing the observed behavior:* First, the behavior in the log is analyzed, extracting state information from it, constructing an automaton representing the observed behavior. This step is explained in detail in Section 5.2.

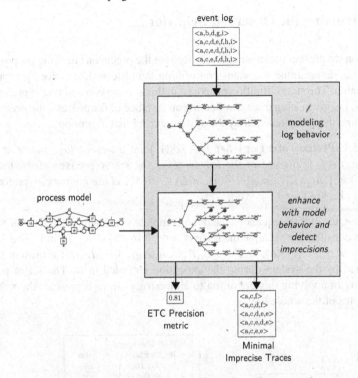

Fig. 5.2: Route map of the precision based on escaping arcs.

2. *Incorporating modeled behavior:* Second, the observed behavior automaton is enhanced by incorporating information about the modeled states. This step does not require an exhaustive exploration of the modeled behavior, but the exploration is restricted to the 'border' between observed and modeled behavior. This step is explained in detail in Section 5.3.
3. *Detecting escaping arcs and estimating precision:* The escaping arcs of a state refer to those points where the model allows more behavior than the one actually recorded in the log. The number of escaping arcs and their location are used to estimate the precision of the whole system in terms of a metric. This step is explained in detail in Section 5.4.
4. *Collecting minimal imprecise traces:* Finally, all the precision problems detected are collected in terms of an event log, describing the minimal traces leading to a precision problem. This step is explained in Section 5.5.

5.2 Constructing the Observed Behavior

In this section we present the first necessary step for the precision checking proposed in this section: determining the states conforming the observed behavior by means of an automaton. The state definition proposed in this section is based on the prefixes of the traces, i.e. given a log trace σ reflecting an instance of the process, the *prefixes* of σ determine the states reached by the system during that execution.

Definition 5.1 (Prefixes of a Log) *Let $L \in \mathscr{B}(A^*)$ be an event log, where $\sigma = \langle a_1, a_2, \ldots, a_n \rangle \in L$ is a trace of the log. $\bullet(\sigma)$ is the set of prefixes of the trace σ, i.e., $\bullet(\sigma) = \{\langle a_1, a_2, \ldots, a_m \rangle | m \leq n\}$. $\bullet(L) \in \mathscr{B}(A^*)$ is the multiset of prefixes of the log L, i.e., $\bullet(L) = \biguplus_{\sigma \in L} \bullet(\sigma)$.*

Notice that, the empty sequence $\langle \rangle$ and the complete sequence are prefixes of any trace. Let us consider for example the trace $\sigma = \langle a, b, d, g, i \rangle$ of log L_1 in Figure 5.3. The prefixes $\bullet(\sigma) = \{\langle \rangle, \langle a \rangle, \langle a, b \rangle, \langle a, b, d \rangle, \langle a, b, d, g \rangle, \langle a, b, d, g, i \rangle\}$ represent the states reached by the system during the execution recorded in σ. The set of prefixes resulting of applying this definition to all the traces in L_1 represents the set of observed states of the whole log.

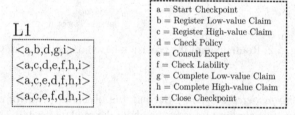

Fig. 5.3: Event log L_1 for a liability insurance claim process.

The state information extracted from a log is used to construct a compact representation in terms of an automaton. The *prefix automaton* contains the states, the transitions between the states, and the weight of a state within the process.

Definition 5.2 (Prefix Automaton of the Log) *Let $L \in \mathscr{B}(A^*)$ be an event log, where A is the set of activities. We define the prefix automaton of the log as an extension of the transition system definition $\mathscr{A}_L = (S, A, \nearrow, \omega, \langle \rangle)$ such that:*

- *the set of states correspond to the set of prefixes of the log, i.e., $S = \{\sigma | \sigma \in \bullet(L)\}$.*
- *the set of labels correspond to the set of activities of the log, i.e., A.*
- *the arcs $\nearrow \subseteq (S \times A \times S)$ define the concatenation between prefixes and activities, i.e., $\nearrow = \{(\sigma, a, \sigma \cdot \langle a \rangle) | \sigma \in S \wedge \sigma \cdot \langle a \rangle \in S\}$.*
- *the function that determines the weight of a state is determined by the number of occurrences of the state in the multiset of prefixes of the log, i.e., $\omega(\sigma) = \bullet(L)(\sigma)$.*

- *the initial state corresponds with the empty prefix* $\langle\rangle$.

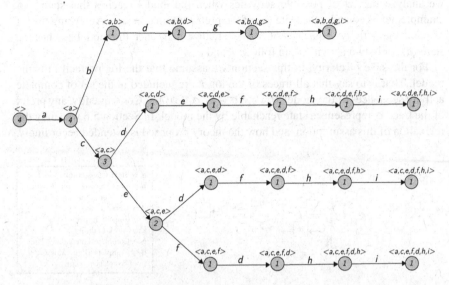

Fig. 5.4: Prefix automaton $\mathscr{A}_{L_1}$ for the event log L_1.

Figure 5.4 illustrates the construction of the prefix automaton for the log L_1. Each prefix of L_1 identifies a state. The number in the states represents the weight function. For instance, the state $\langle a \rangle$ has a weight of $\omega(\langle a \rangle) = 4$ because it appears four times in $\bullet(L)$. On the other hand, $\langle a, b \rangle$ appears only once, i.e., $\omega(\langle a, b \rangle) = 1$. The initial state corresponds with the prefix $\langle\rangle$.

In [14], the authors proposed a configurable approach to construct a transition system from an event log. The definition of state and the events considered to build a state can be parametrized depending on the abstraction level desired. In particular, the parameters *past*, *sequence*, and *no horizon* will result in a transition system with the same characteristics as the prefix automaton proposed in this section. In Chapter 10 we consider the use of other values for the parameters and their effects on the precision checking.

5.3 Incorporating Modeled Behavior

In this section we use the prefix automaton representing the observed behavior to restrict the exploration of the modeled behavior. Let us consider, for example, the model SN_1 in Figure 5.5, presented in terms of a system net, with [*start*] and [*end*]

as initial and final markings respectively.[1] The system net is a possible model for
the insurance process observed in the log L_1. Given a state of the prefix automaton,
we analyze the set of possible activities when the model reaches that state. For
example, let us consider the state $\langle a \rangle$ of the prefix automaton $\mathscr{A}_{L_1}$. Analyzing the set
of complete activity sequences of SN_1 we realize that, once the model reaches the
state $\langle a \rangle$, only two activities can follow: b and c.

For the sake of clarity, in this section we assume that the log perfectly fits the
model. That is to say that all traces of the log L are included in the set of complete
activity sequences of the model M, i.e. $\forall \sigma \in L : \sigma \in \phi(M)$. Consequently, any prefix
of the trace σ represents a state reachable by the model. In Section 5.6 we study the
relaxation of this assumption, and how the theory proposed is extended accordingly.

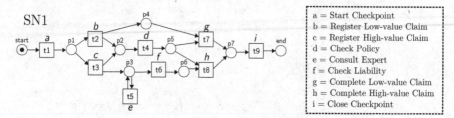

Fig. 5.5: System net SN_1 modeling the liability insurance claim process.

Given a prefix automaton from the log, we analyze the activities allowed by the
model in each log state, and we incorporate new states denoting those activities
modeled but never observed in the log. Let us consider for example the automaton
$\mathscr{A}_{L_1 SN_1}$ in Figure 5.6, result of enhancing $\mathscr{A}_{L_1}$ with modeled behavior of SN_1. The
states in white represent states that are in both the log and model behavior. The
colored states represent the new states introduced, that belong only to modeled behavior but never observed in the log. The weightf of these new states is 0, denoting
that they are not observed in the log. For example, the state $\langle a, c \rangle$ represents a state
in both the log L_1 and the model SN_1, but $\langle a, c, f \rangle$ is a state only of the model (i.e.,
$\langle a, c, f \rangle$ is not a prefix of L_1).

Formally, the enhancement of the observed prefix automaton with modeled behavior is defined as follows:

**Definition 5.3 (Prefix Automaton of the Observed Behavior Enhanced with the
Modeled Behavior)** *Let $L \in \mathscr{B}(A^*)$ be an event log, where A is a set of activities. Let
M be a process model, where $\phi(M)$ is the set of complete activity sequences of M.
Similar to $\bullet(L)$, $\bullet(\phi(M))$ represents the prefixes of the complete activity sequences
of M, i.e., $\bullet(\phi(M)) = \biguplus_{\sigma \in \phi(M)} \bullet(\sigma)$. We define the prefix automaton of the observed
behavior enhanced with modeled behavior $\mathscr{A}_{LM} = (S, A, \nearrow, \omega, \langle \rangle)$ such that:*

---

[1] Notice that we use system nets in this section for illustrative purposes only, while the theory
refers to any process model.

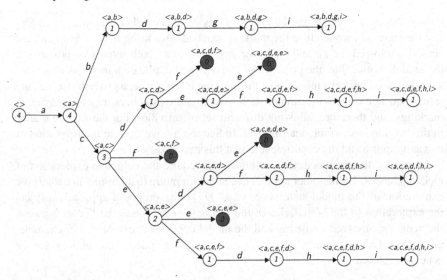

Fig. 5.6: Prefix automaton $\mathscr{A}_{L_1 SN_1}$ of the event log L_1 enhanced with the modeled behavior of the system net SN_1.

- Let $S_{LM} = \{\sigma | \sigma \in \bullet(L) \cap \bullet(\phi(M))\}$ be the states that are in both the observed and modeled behavior. Let $S_M = \{\sigma | \sigma \notin \bullet(L) \wedge \sigma \in \bullet(\phi(M))\}$ be states that are in the modeled behavior but not in the observed behavior[2]. The set of states S of the enhanced automaton is the union $S = S_{LM} \cup S_E$, where $S_E \subseteq S_M$ are the states only in the modeled behavior that come preceded by a state in both behaviors, i.e., $S_E = \{\sigma \cdot \langle a \rangle | \sigma \in S_{LM} \wedge \sigma \cdot \langle a \rangle \in S_M\}$.
- The set of labels correspond to the set of activities of the process, i.e., A.
- The arcs $\nearrow \subseteq (S \times A \times S)$ define the concatenation between states and activities, i.e., $\nearrow = \{(\sigma, a, \sigma \cdot \langle a \rangle) | \sigma \in S \wedge \sigma \cdot \langle a \rangle \in S\}$.
- The function that determines the weight of a state is determined by the number of occurrences of the state in the multiset of prefixes of the log P with $P = \bullet(L)$, or 0 if the state does not appear on the log:

$$\omega(\sigma) = \begin{cases} P(\sigma) & \text{if } \sigma \in S_{LM} \\ 0 & \text{if } \sigma \in S_E \end{cases}$$

- The initial state corresponds with the empty prefix $\langle \rangle$.

Notice that, although the definition of the enhanced prefix automaton considers the set of complete activity sequences $\phi(M)$ of a model M, in practice the approach proposed does not require computing all the sequences in advanced. In the previous example, for the system net SN_1 the set of complete activity sequences is infinite due the transition t_5. Instead, in executable process models such as Petri nets or BPMN,

---

[2] $S_L = \{\sigma | \sigma \in \bullet(L) \wedge \sigma \notin \bullet(\phi(M))\}$ is not possible because we assume perfect fitness (cf. Section 5.6.)

the sequences can be constructed progressively on demand. For example, given SN_1 and the state $\langle a \rangle$, we analyze the marking reached after firing t_1 (i.e., $[p_1]$) and the activities allowed (i.e., b and c), being $\langle a,b \rangle$ and $\langle a,c \rangle$ both sequences possible in the model. Notice that, the presence of invisible and duplicate transitions may arise some potential indeterminism about the marking reached after a given prefix, i.e., the same sequence of activities may match several tasks sequences, reaching different markings, and therefore, allowing different set of activities. For the sake of clarity in this section we assume determinism. In Section 5.6 we explore the relaxation of this assumption and the consequences that this may produce.

Moreover, the proposed approach does not require the complete exploration of model behavior. Unlike other approaches in the literature that require an exhaustive exploration of the model state space (e.g., [77]), the proposed approach restricts the exploration to the boundaries of the log state space. Only the border between the behavior observed in the log and the model behavior is explored. For example, given L_1 and SN_1, the trace $\langle a,c,d,e,e,e,e,e,f,h,i \rangle$ included in the model behavior is never explored.

5.4 Detecting Escaping Arcs and Evaluating Precision

Given an event log L and process model M, the prefix automaton $\mathscr{A}_{LM}$ contains the juxtaposition of both observed and modeled behaviors, and the border defined between them. Each state of the automaton represents a state reached by the system during the execution recorded in the log. The precision checking approach proposed in this chapter bases its estimation on comparing, for each one of these states, the activities allowed by the model and the activities that where recorded in the log.

Definition 5.4 (Observed and Modeled Arcs) *Let $\mathscr{A}_{LM} = (S, A, \nearrow, \omega, \langle \rangle)$ be the prefix automaton of the log L enhanced with the behavior of the model M. Let S_{LM} be the states that are in both observed and modeled behavior, and let S_M be the states only in the modeled behavior. Let $\sigma \in S_{LM}$ be a state of the automaton. $obs(\sigma) = \{(\sigma, a, \sigma') \in \nearrow \,|\, \sigma \in S_{LM} \wedge \sigma' \in S_{LM}\}$ represent those arcs whose activities were executed and consequently recorded on the log L when the system was on the state σ. $mod(\sigma) = \{(\sigma, a, \sigma') \in \nearrow \,|\, \sigma \in S_{LM} \wedge \sigma' \in S_{LM} \cup S_M\}$ represent those arcs whose activities are* modeled *and consequently allowed by the model M when the system was on the state σ. Similarly, we refer as observed/modeled states of σ those states reached through an observed/modeled arc from σ, and observed/modeled activities of σ those activities used in the observed/modeled arcs from σ.*

For example, let us consider the state $\langle a,c \rangle$ of the automaton $\mathscr{A}_{L_1 SN_1}$. The activities modeled by the model SN_1 in that state are d, e and f. On the other hand, the activities observed in the same state are only d and e. Notice that in this section we are under the assumption that the log perfectly fits the model. Therefore, $observed(\sigma) \subseteq modeled(\sigma)$ for all states $\sigma \in S_{LM}$.

An *escaping arc* denotes a point where the behavior of the process model allows more than what has been actually observed in the log, i.e., an arc that *escapes* from the log behavior.

Definition 5.5 (Escaping Arcs) *Let $\mathscr{A}_{LM} = (S, A, \nearrow, \omega, \langle \rangle)$ be the prefix automaton of the log L enhanced with the behavior of the model M, where $\sigma \in S_{LM}$. The set of escaping arcs of the state σ is determined by the difference between the arcs modeled and the arcs allowed on the state, i.e., $esc(\sigma) = mod(\sigma) \setminus obs(\sigma)$. Similarly, we refer to the activities used to escape and the states reached, as escaping activities and escaping states, respectively. The set of all escaping arcs of the automaton is denoted as $esc(\mathscr{A}_{LM})$.*

Following with the example state $\langle a, c \rangle$ of the automaton in $\mathscr{A}_{L_1 SN_1}$, there is only one escaping activity in that state: f. In other words, when the system was in the state $\langle a, c \rangle$, all activities allowed by the model in that point have been observed, except f. Notice that, by construction of the automaton, the escaping activities are considered *globally*, i.e., all the traces in the log are considered as a whole to compute the set of reflected activities, instead of analyzing trace by trace independently. For example, given the state $\langle a, c \rangle$ in $\mathscr{A}_{L_1 SN_1}$, the activity d is reflected in the second trace of L_1, and the activity e is reflected in the third and forth traces of the log.

In our view, a precise model is one that does not contain escaping arcs, i.e., for each state it only models the behavior observed on the log. A model where almost all the behavior allowed represents an escaping opportunity must be considered highly imprecise. In that sense, we define a metric –*ETC Precision*– designed to measure the precision between a model and a log, based on the escaping arcs. On the one hand, the metric quantifies the degree of escaping arcs of the automaton. This value is weighted according to the weight of the state where each escaping arc is located, i.e., escaping arcs in more frequent states have more impact in the metric that those who appear in infrequent and barely used parts of the process. On the other hand, the metric measures the modeled behavior on the automaton, weighted also according to the weights of the states. The metric defines the precision between a log and a model as the relation between the escaping behavior versus the modeled behavior.

Definition 5.6 (ETC Precision) *Let $\mathscr{A}_{LM} = (S, A, \nearrow, \omega, \langle \rangle)$ be the prefix automaton of the log L enhanced with the behavior of the model M. The metric ETC Precision estimates the precision of the system comparing, for each state in S_{LM}, the number of escaping arcs with the number of modeled arcs. The numbers are weighted according to the importance of the state. Formally:*

$$etc_p(\mathscr{A}_{LM}) = 1 - \frac{\sum_{\sigma \in S_{LM}} \omega(\sigma) \cdot |esc(\sigma)|}{\sum_{\sigma \in S_{LM}} \omega(\sigma) \cdot |mod(\sigma)|}$$

Let us consider, for example, the automaton $\mathscr{A}_{L_1 SN_1}$. The automaton contains 21 states in S_{LM}, denoted in white. For each one of those states we compute the number of escaping arcs and the number of modeled arcs, and we weight them according to the weight of each state.

$$etc_p(\mathscr{A}_{L_1 SN_1}) = 1 -$$

$$\frac{4 \cdot 0 + 4 \cdot 0 + 1 \cdot 0 + 1 \cdot 0 + 1 \cdot 0 + 1 \cdot 0 + 3 \cdot 1 + 1 \cdot 1 + 1 \cdot 1 + \cdots}{4 \cdot 1 + 4 \cdot 2 + 1 \cdot 1 + 1 \cdot 1 + 1 \cdot 1 + 1 \cdot 0 + 3 \cdot 3 + 1 \cdot 2 + 1 \cdot 2 + \cdots}$$

$$\frac{\cdots 1 \cdot 0 + 1 \cdot 0 + 1 \cdot 0 + 2 \cdot 1 + 1 \cdot 1 + 1 \cdot 0 + 1 \cdot 0 + 1 \cdot 0 + 1 \cdot 0 + \cdots}{\cdots 1 \cdot 1 + 1 \cdot 1 + 1 \cdot 0 + 2 \cdot 3 + 1 \cdot 2 + 1 \cdot 1 + 1 \cdot 1 + 1 \cdot 0 + 1 \cdot 1 + \cdots}$$

$$\frac{\cdots 1 \cdot 0 + 1 \cdot 0 + 1 \cdot 0}{\cdots 1 \cdot 1 + 1 \cdot 1 + 1 \cdot 0} = 1 - \frac{8}{43} = 1 - 0.19 = 0.81$$

The etc_p value for the automaton $\mathscr{A}_{L_1 SN_1}$ between L_1 and SN_1 is 0.81, denoting a moderate precision degree of 81%.

Taking a look at the literature one can see that the intuitive notion of precision is difficult to capture in a metric. Comparing a process model and an even log always allows for different interpretations and a wide range of metrics can defined. Facing so much uncertainty it is wise to impose some requirements to ensure the usefulness of any measure proposed. In [74] the authors present a list, based on [52], with five properties any conformance metric proposed should satisfy: *validity, stability, analyzability, reproducibility* and *localizability*. In the following part we analyze those properties and we provide a brief justification on how ETC Precision metric fulfills them:

- **Validity** *Validity means that the measure and the property to measure must be sufficiently correlated with each other.*
  As it has been motivated, there is a direct relation between the precision of a system and the escaping arcs detected, justifying its validity. An increase in the precision degree is reflected in the number and importance of the escaping arcs, and thus, producing an increment on the metric value.
- **Stability** *Stability means that the measure should be stable against manipulations of minor significance, i.e., be as little as possible affected by properties that are not measured.*
  The approach proposed is defined at a activity level, i.e., the observed behavior is compared with the modeled behavior independently from the structural properties of the models. Two models with different structure but modeling the same behavior will result in the same metric value. This makes it even possible to compare models defined using different notations, e.g., Petri nets and BPMN. The metric is defined such that the context of the precision problem is taken into account, but not the position, i.e., two states with the same weight will have the same importance in the metric no matter where they are located. Notice that, although the metric is defined to measure the precision dimension independently, possible problems and corrections in the fitness dimension may affect the stability of the precision results (cf. Section 5.6).
- **Analyzability** *Analyzability, in general, relates to the properties of the measured values (e.g., whether they can be statistically evaluated). In the remainder, the emphasis is on the requirement that the measured values should be distributed between 0 and 1, with 1 being the best and 0 being the worst value.*
  Notice that by definition $esc(\sigma) \subseteq mod(\sigma)$, resulting in a metric that range from 0 to 1. Therefore, on the one hand an optimal value for precision is defined,

i.e, 1 denotes that the observed behavior is precisely modeled. This is especially important as a stop condition in the context of an iterative approach looking for appropriate process models, such as genetic mining [56, 34], but also for a human analyst as it indicates that there is no better solution available. Notice that, to achieve a value of 1 it is not necessary to have all the modeled behavior observed in a single trace, i.e., the precision is considered globally, taking all the observed traces as a whole. Finally, the fact that the metric is normalized by the degree of allowed behavior in each state makes it possible to be used for comparing different pairs of model-log, even if they refer to different processes.

- **Reproducibility** *Reproducibility means that the measure should be independent of subjective influence, i.e., it requires a precise definition of its formation.*
  The definition of the metric proposed is solid and formal, and there is no room for subjectivity. The same experiment can be reproduced several times and it will always output the same result. However, notice that the relation between activity sequences and tasks sequences of a process model may arise possible non-deterministic situations, requiring further assumptions in order to preserve the reproducibility of the approach (cf. Section 5.6).

- **Localizability** *Localizability means that the system of measurement forming the metric should be able to locate those parts in the analyzed object that lack certain desirable (i.e., the measured) properties.*
  It is very important that a precision problem is not only reflected by the measured value but can also be located. In that sense, the escaping arcs captured in the automaton describe perfectly where the precision problems are, making it possible for the business analyst to identify potential points of improvement. Additionally, in Section 5.5 we provide an additional mechanism to collect all those precision problems for a deeper analysis.

5.5 Minimal Imprecise Traces

The ETC Precision metric presented in the previous section provides a numeric measurement on the precision of the system. This value may be useful to measure the precision of several alternative models describing the same observed behavior, or to establish when a model becomes obsolete to represent an evolving process. However, in order to fully understand the causes of the precision problems, an analyst needs to be able to access the exact points of mismatch between the observed and modeled behaviors. The prefix automaton and its detected escaping arcs provide this information, and may be used to guide a deeper analysis into the model and the log to understand their discrepancy. Some of the escaping arcs may represent meaningful abstractions that arise in the model and therefore no further action is required. Others, however, may suggest situations for which future actions over the process need to be carried out.

Additionally to the escaping arcs detected on the automaton, in this section our purpose is to collect all the precision anomalies in terms of logs and traces to be used later on for its analysis. Each escaping arc is represented by one *minimal imprecise trace (mit)*, a sequence containing the minimal behavior observed until the escaping arc was available. All the minimal imprecise traces compose the *minimal imprecise log*.

Definition 5.7 (Minimal Imprecise Traces and Log) *Let* $\mathscr{A}_{LM} = (S, A, \nearrow, \omega, \langle\rangle)$ *be the prefix automaton of the log L enhanced with the behavior of the model M. Let* $esc(\mathscr{A}_{LM})$ *define all the escaping arcs of the automaton. Given the escaping arc* $(\sigma, a, \sigma') \in esc(\mathscr{A}_{LM})$, *its minimal imprecise trace is defined as* $mit((\sigma, a, \sigma')) = \sigma'$. *The set of all minimal imprecise traces defines the* minimal imprecise log, *i.e.,* $mil(\mathscr{A}_{LM}) = \bigcup_{i \in escape(\mathscr{A}_{LM})} mit(i)$.

In $\mathscr{A}_{L_1 SN_1}$, there are five escaping arcs in the automaton, and thus, five are the minimal imprecise traces conforming the minimal imprecise log, shown in Figure 5.7. Notice that, by definition, all minimal imprecise traces fulfill a minimality criterion, i.e., they represent the minimal behavior before the deviation. In other words, all elements in the trace except the last one represent a behavior observed in the log, and the last one is the activity allowed by the model but not observed.

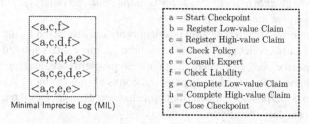

Minimal Imprecise Log (MIL)

Fig. 5.7: Minimal Imprecise Log (MIL) for the automaton $\mathscr{A}_{L_1 SN_1}$.

The representation of the imprecisions in terms of a minimal imprecise log opens the possibility to different analysis and uses. For example, all techniques and tools based in analyzing event logs can be used to gain insights into the precision problems. That includes most of the approaches in the field of process mining, e.g., discovery algorithms can be used to derive a model that represents the imprecise behavior. Some of the minimal imprecise traces can be used for *process reparation*, correcting the model to represent more precisely the observed behavior, e.g., transforming pairs of concurrent events in the model to ordered events [57]. Another option is to use the minimal imprecise log as input for the *supervisory control* theory [71], i.e., synthesizing a supervisor (i.e., another model synchronized with the original model) that restricts the behavior of original system such the imprecisions never occur. Finally, similar to [45, 88, 30], the minimal imprecise trace can be considered as negative examples, and can be used to enrich the original log in order to discover a more precise model.

5.6 Limitations and Extensions

For the sake of clarity, the approach presented made some assumptions. In this section we discuss the effects of relaxing those assumptions, addressed in later chapters of the book. In particular, we focus on the consequences of unfitting traces and the indeterminism between activity sequences and tasks sequences.

5.6.1 Unfitting Scenario

This chapter has presented the precision checking based on escaping arcs assuming a perfectly fitting log. In other words, each trace on a log L is included in the set of complete activity sequences of the model M. Consequently, any prefix in the log traces is a prefix of the sequences of the model, and therefore, $observed(\sigma) \subseteq modeled(\sigma)$ for all states $\sigma \in S_{LM}$.

However, this is a strong assumption, especially in real-life scenarios, where misrecorded events and exceptional executions of the process exist. For such cases, we define $S_L = \{\sigma | \sigma \in \bullet(L) \wedge \sigma \notin \bullet(\phi(M))\}$ as the set of states in the observed behavior but not modeled. Those states represent situations difficult to interpret, where the observed behavior is not aligned within the domain of the model [17]. In other words, the model is not able to determine the state of the system given the observed elements.

Given the rare nature of these situations, a possible strategy is to consider the fitting part of the traces for the computation of the precision. In other words, given an unfitting trace $\sigma = \sigma' \cdot \langle a \rangle \cdot \sigma''$ where $\sigma' \in S_{LM}$ and $\sigma' \cdot \langle a \rangle \in S_L$, only σ' is used to compute precision. For example, let us consider the model M of Figure 5.8 and the log $L = [\langle a,b,d,f \rangle^{100}, \langle a,c,e,f \rangle^{100}, \langle a,b,c,d,f \rangle]$. The observed behavior used to compute the precision automaton is composed by the two first traces (100 instances of each) and the fragment $\langle a,b \rangle$ of the third trace.

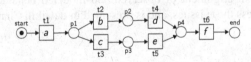

Fig. 5.8: Model to illustrate the problems of unfitting scenarios.

Notice that the current definition of observed, modeled and escaping arcs presented in Definition 5.4 and 5.5 already satisfies that assumption, considering observed arcs those arcs in the behavior of both the log and the model. Moreover, the *ETC Precision* proposed in Definition 5.6, based on observed and escaping arcs, also satisfies the assumption. If the escaping arcs suggest situations with precision problems, we define the equivalent *fitness escaping arcs* as the exact points where the observed behavior deviates from the model. In these cases, the fitness escaping arcs do not provide information about the precision dimension, but simple information about the fitness of the system. Moreover, similar to the minimal imprecise trace (cf. Definition 5.7), we define the *minimal unfitting trace* as the trace fragment reaching that fitness escaping arcs. We define the *minimal unfitting log* as the collection of those unfitting traces in terms of a log. For example, $L = [\langle a,b,d,f \rangle^{100}, \langle a,c,e,f \rangle^{100}, \langle a,b,c,d,f \rangle]$ has only one fitness escaping arc, and therefore the minimal unfitting log is composed only by the trace $\langle a,b,c \rangle$.

The strategy of considering the fitting part of the traces to compute precision is only appropriate for those cases where the unfitting behavior represent a minor part of the whole observed behavior and its effect on the metric is negligible. However, it has several disadvantages and limitations. First, the precision metric is affected by the position of the fitness problem, creating a dependency between both dimensions not desired [21, 20]. In the worst case scenario, when the fitness mismatch is at the beginning of the trace, the whole trace is discarded. For example, $L_1 = [\langle a,a,b,d,f \rangle, \langle a,a,c,e,f \rangle]$ and $L_2 = [\langle a,b,d,f,f \rangle, \langle a,c,e,f,f \rangle]$ are two logs for the model M where a or f is recorded twice by mistake. In L_1, only $\langle a \rangle$ is considered to measure precision (the rest of both traces is discarded). This results in a low precision value of 0.3. On the other hand, in L_2, $\langle a,b,d,f \rangle$ and $\langle a,c,e,f \rangle$ are considered (only the last f is discarded), and this result in a perfect precision of 1.

The second limitation is that noisy unfitting behavior distort the precision analysis, "covering" arcs that otherwise would be considered escaping. For example, let us consider the log $L = [\langle a,b,d,f \rangle^{1000}, \langle a,c,b,d,f \rangle]$. In this log, the second trace, that could be considered noisy, covers the escaping arc $\langle a,c \rangle$, denoting that the lower path of the model is never used. Notice that the effect on the metric from a escaping arc (i.e., c) in a state $\langle a \rangle$ with a weight of 1000 is much more than the effect of the escaping arc e in a state $\langle a,c \rangle$ with a weight of 1.

This issue is related with the third limitation: the fitness anomalies are not addressed globally, but locally. For example, given the trace $\langle a,c,b,d,f \rangle$ in the aforementioned example, we assume that deviation is produced after $\langle a,c \rangle$. However, a global vision would consider a more logical interpretation where c is already a deviation, denoting a misrecorded event on a perfectly fitting trace $\langle a,b,d,f \rangle$. The application of global alignment of observed and modeled behavior for precision checking will address these issues, and is presented in detail in Chapter 9.

5.6.2 Indeterministic Scenario

The second assumption made for the sake of clarity during this chapter is the deterministic relation between activity sequences and tasks sequences of a model. In other words, given a sequence of activities, there is only one sequence of tasks associated with it. However, this is not always the case. Let us consider the model in Figure 5.9. Given the activity sequence $\langle a,b \rangle$, there are two transition sequences resulting in that sequence: $\langle t_1,t_2 \rangle$ and $\langle t_1,t_3 \rangle$. A similar situation may occur when the model contains invisible tasks.

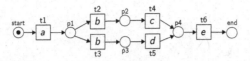

Fig. 5.9: Model to illustrate the problems of indeterministic scenarios.

There are different strategies to deal with indeterminism. One option is to consider and explore all possible scenarios. However, this solution could lead to a state-space explosion, making it only suitable for small and simple cases [77]. Other options include the use of heuristics, e.g., a random selection, or more elaborate look-ahead heuristics to determine the most plausible option. The heuristic solution contrasts with the possibility of using alignment algorithms to determine the global optimal sequence of tasks for a given activity sequence. This possibility is explored in detail in the Chapter 9.

5.7 Summary

Given an event log and a process model, the notion of escaping arcs offers an efficient alternative to analyze the precision dimension. This chapter described the definition of the observed behavior in terms of an automaton, and how that automaton is enhanced to incorporate information about the modeled behavior. It showed how the enhanced automaton is used to detect the escaping arcs. The escaping arcs, and their position within the process, are the bases for the proposed precision metric.

Chapter 6
Evaluating Precision in Practice

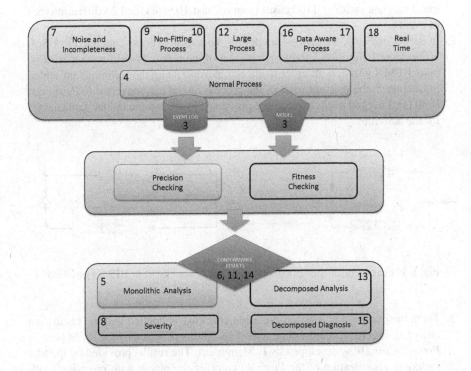

Previous chapters motivated the need for precision checking and introduced a precision measure based on escaping arcs. This chapter illustrates the evaluation of precision in practice using the University case presented in Chapter 1. Additionally, the chapter presents empirical results to illustrate the characteristics of the approach. Later chapters consider variants and extensions of the procedure to adapt it to particular conditions.

© Springer International Publishing AG 2016
J. Munoz-Gama: Conf. Check. ... in Process Mining, LNBIP 270, pp. 55–59, 2016.
DOI: 10.1007/978-3-319-49451-7_6

6.1 The University Case: The Appeals Process

In Chapter 1, the University Case is presented: a university is planning on changing its obsolete system which manages academic processes to a new system called
USoft. For that, the university compares the event data generated by the legacy system with the process models of USoft in order to analyze the precision.

In this section we describe the precision analysis performed for the *appeals process*, the process that concerns those cases where the student needs to be expelled
from the university, for low academic results, severe ethics violations, or other circumstances.

The precision analysis is performed as follows:

1. First, the data recorded by the legacy system is processed. Several tables of the
   system database containing the information regarding the appeals process are
   identified, and the information is extracted and consolidated in terms of an event
   log. That data includes 1100 cases from the last 10 years, and 23 different types
   of events.
2. In the second step, the process model for the appeal process contained in USoft
   is considered. Figure 6.1 shows a fragment of the USoft model for such process,
   in its BPMN notation. In order to check precision using the technique presented
   in Chapter 5, the relevant parts of the process model are converted to Petri nets.
   Both Petri net and event logs are pre-processed in order to guarantee same names
   for the activities.

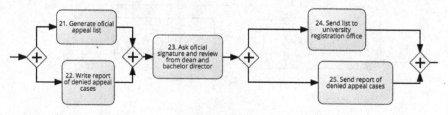

Fig. 6.1: Fragment of the appeals process model in USoft in BPMN notation.

3. Then, the escaping arcs based precision checking is applied and the results are
   analyzed. Figure 6.2 show a fragment of the results provided by the ProM plug-in
   Precision for DPN, developed by F. Mannhardt. The results provided by the tool
   include a visualization of the Petri net coloring the places with precision problems (top right), the minimal imprecise traces denoting the comparison between
   observed and possible arcs (top left), and a summary of the precision and arcs,
   including the observed and possible arcs for each place.
4. After analyzing the results, the university concludes that the USoft model for the
   appellation process is precise enough for their requirements (0.778). Moreover,
   the university identifies two precision problems from the results. First, the model
   allows for t21 and t22 in parallel. However, the log reflects a sequential order

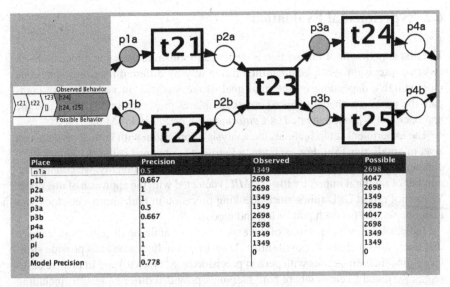

Fig. 6.2: Fragment of the precision results provided by *Precision for DPN* tool.

between t21 and t22. Although the tool identifies this problem, the university considers this imprecision as irrelevant: the list is generated automatically and it takes less than a second to complete task t21, but writing the report is a time consuming task (t22). Therefore it is normal that the sequence t22-t21 is never observed in the log. Second, the approach detects that t25 can be executed before t24. This is a violation of the university policy (the resolution needs to be registered first before sending the reports), and consequently the USoft model needs to be adapted before its implementation.

5. Finally, the USoft appeals model is adapted to correct the precision problems between the real process and the model identified on the conformance analysis. Figure 6.3 shows a fragment of the adapted USoft model.

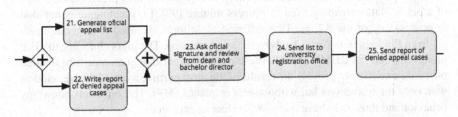

Fig. 6.3: Fragment of the *adapted* appeals process model in USoft in BPMN notation.

6.2 Experimental Evaluation

The first experiment of this section is designed to illustrate the same dimension – in this case precision – can be quantified differently by different measures, each one more suitable depending on the final goal of the analysis. In particular, we compare the precision checking based on escaping arcs presented in this book, and the approach in [77], implemented as *Conformance Checker* in ProM 5.2.

The experiment setup includes the analysis of precision with both approaches on a set of small logs [59]. For each log, a Petri net is obtained using ILP miner [93], a miner that provides perfectly fitting process models. Additionally, we counted the number of minimal imprecise traces ($|MIL|$) detected with the approach of this book. We do not report CPU times since checking precision in both approaches took less than one second for each pair of log and model.

Table 6.1a shows the results of the experiment. Examining the table, one can see that the results include all possibilities: 1) cases where both measures provide similar results, including cases with perfect precision (e.g., GFA6NTC) and imprecise cases (e.g., GFA5), and 2) cases where both measures provided different results, including cases where [77] provides a higher measurement (e.g., GFAL1), and cases where the approach of this book is higher (e.g., GFBN2). The reason for such differences depends on the techniques used to measure precision: in general, [77] highly penalize a general discrepancy between log tasks relations and model tasks relations, while precision based on escaping arcs only penalize the few situations where the escaping arcs appear in the log.

The first conclusion arised from the experiment is that the same dimension could be measured in different ways. Based on that, one may decide to use [77] for a more holistic vision of the precision, while the approach of this book could be used to detect the specific precision discrepancies. However, as the next experiment illustrates, for non trivial examples, the only alternative is the precision based on escaping arcs.

The purpose of the second experiment in this section is to illustrate the applicability of the precision checking based on escaping arcs in large processes that other approaches are not able to handle. In particular, we compare with the approach in [77], implemented as *Conformance Checker* in ProM 5.2.

The experiment setup includes the analysis of escaping arcs precision on 8 logs of a public dataset widely used in process mining [93]. [77] technique is not able to process any of the 8 logs. Two different algorithms are used to generate models from the logs in order to analyze the conformance. The algorithms are the ILP miner [93] and the RBMiner [84]. These are two miner that guarantee fitness value one. Additionally, we provide the results of the most permissive models, i.e., models with only the transitions but without arcs or places (M_T). These models allow any behavior and thus, they have a low etc_p value, as expected.

Table 6.1b shows the results of the experiment. The results include the precision value, number of $|MIL|$ and CPU time in seconds for checking precision on the models obtained by the ILP miner, the RBMiner, and the only-transitions models. For each one of the aN benchmarks, N represents the number of tasks in the log, while the _1 and _5 suffixes denote its size: 100 and 900 traces, respectively. The *t*32 has 200 (_1) and 1800 (_5) traces. The pair of CPU times reported denote the computation of etc_p without or with the collection of $|MIL|$ (in parenthesis).

A first conclusion on the results is the capability of the approach to handle large benchmarks in reasonable CPU time, even for the prototype implementation carried out. A second observation is that as the number of tasks increases, precision in the model drops as the discovery algorithms can no longer precisely capture the complex relations between tasks and have to derive less precise relations between them.

Benchmark	[77]	etc_p	$|MIL|$	Benchmark	[77]	etc_p	$|MIL|$
GFA6NTC	1.00	1.00	0	GFl2lOpt	1.00	0.85	7
GFA7	1.00	1.00	0	GFAL2	0.86	0.90	391
GFA8	1.00	1.00	0	GFDrivers	0.78	0.89	2
GFA12	1.00	1.00	0	GFBN3	0.71	0.88	181
GFChoice	1.00	1.00	0	GFBN2	0.59	0.96	19
GFBN1	1.00	1.00	0	GFA5	0.50	0.57	35
GFParallel5	1.00	0.99	11	GFl2l	0.47	0.75	11
GFAL1	1.00	0.88	251	GFl2lSkip	0.30	0.74	10

(a)

		M_T	Parikh				RBMiner			
Benchmark	$|TS|$	etc_p	$|P|$	$|T|$	etc_p	$|MIL|$	CPU	$|P|$	$|T|$	etc_p	$|MIL|$	CPU
a22f0n00_1	1309	0.06	19	22	0.63	1490	0(0)	19	22	0.63	1490	0(0)
a22f0n00_5	9867	0.07	19	22	0.73	9654	0(3)	19	22	0.73	9654	0(4)
a32f0n00_1	2011	0.04	31	32	0.52	2945	0(0)	32	32	0.52	2944	0(1)
a32f0n00_5	16921	0.05	31	32	0.59	22750	2(10)	31	32	0.59	22750	2(11)
a42f0n00_1	2865	0.03	44	42	0.35	7761	0(2)	52	42	0.37	7228	0(2)
a42f0n00_5	24366	0.04	44	42	0.42	60042	5(28)	46	42	0.42	60040	6(29)
t32f0n00_1	7717	0.03	30	33	0.37	15064	1(15)	31	33	0.37	15062	1(12)
t32f0n00_5	64829	0.04	30	33	0.39	125429	9(154)	30	33	0.39	125429	8(160)

(b)

Table 6.1: (a) Comparison of the precision results between the proposed approach and approach in [77] for small examples. (b) Comparison of precision results between models obtained by three discovery algorithms, for large examples where [77] was not able to finish.

Chapter 7
Handling Noise and Incompleteness

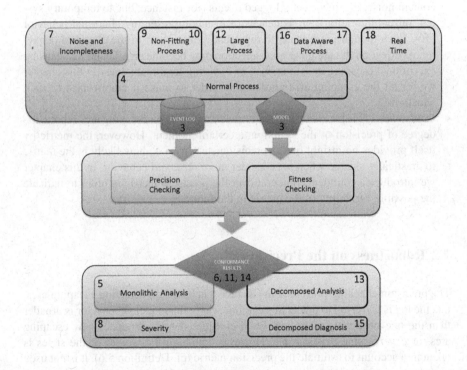

In the previous chapters we presented a precision checking approach for noise-less data. However, noise is present in most real-life cases. Moreover, the behavior reflected in the log can only be considered as an example of the real process behavior. This chapter presents the adaptation of the precision presented in the first chapters of this book in order to be more robust against noise, and it provides a

© Springer International Publishing AG 2016
J. Munoz-Gama: Conf. Check. ... in Process Mining, LNBIP 270, pp. 61–73, 2016.
DOI: 10.1007/978-3-319-49451-7_7

technique to assess the confidence on the variability of measure in the future. In the next chapter we will assess the severity of the precision issues detected.

7.1 Introduction

In the previous chapter we introduced precision checking based on escaping arcs. Given a log and a model, we use the behavior observed in the log to traverse the modeled behavior, detecting the escaping arcs and estimating the precision of a system. In other words, it allows to know *how precise is our model describing the behavior observed in the log*. In this chapter, we go a step further, and we extend the escaping arcs analysis to address the following points:

- *Robustness on the escaping arcs*. Experience has proved that most real-life logs contain noise, e.g, incorrectly logged events (for instance, due to temporary system misconfiguration), or sequences of events representing abnormal behavior [56]. Even though noise tends to represent a minor part of the observed behavior, its effect on the escaping arcs detection may become significant, i.e., causing exceptional escaping arcs to appear, or covering legitimate ones. In this chapter we revisit the escaping arcs approach in order to make it more robust to such situations.
- *Confidence of the precision metric*. A checking based on escaping arcs reports the degree of precision of the system at a certain moment. However, the metric by itself provides no insight on the precision stability, i.e., how likely is the metric to drastically change when we consider more observed behavior. In this chapter we introduce a confidence interval over the precision metric in order to indicate the possible variability of the metric in the future.

7.2 Robustness on the Precision

The precision checking presented in Chapter 5 is based on detecting escaping arcs, i.e., the log is traversed to detect those points where the modeled behavior is broader than the one observed on the log. Each considered trace produces new escaping arcs, or *covers* some existing one. However, although the weight of the states is taken into account to estimate the precision metric (cf. Definition 5.6), it is not used to determine what represents or not an escaping arc. In other words, infrequent or noise behavior in the log may impact on the escaping arcs and the precision metric, covering possible escaping arcs and creating new ones.

Let us consider the insurance process in Chapter 5 and use it as a running example to illustrate the concepts introduced in this chapter. The system net of Figure 7.1 represents a model M for the insurance process, and the log L in Table 7.1 is a reflection of the process execution. Each row of the log represents a trace and the

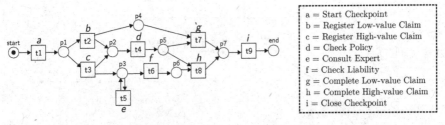

Fig. 7.1: Model M for the insurance process, used as running example to illustrate the concepts of this chapter.

Frequency	Trace
1435	$\sigma_1 = \langle a,b,d,g,i \rangle$
946	$\sigma_2 = \langle a,c,d,e,f,h,i \rangle$
764	$\sigma_3 = \langle a,c,e,d,f,h,i \rangle$
54	$\sigma_4 = \langle a,c,e,f,d,h,i \rangle$
1	$\sigma_5 = \langle a,c,d,e,e,f,h,i \rangle$

Table 7.1: Event log L for the insurance process, used as running example to illustrate the concepts of this chapter.

number of occurrences of the trace in the log. Therefore, L captures 3200 executions of the process, following 5 different paths.

Figure 7.2 shows the prefix automaton of L and M as it is presented in Chapter 5. Looking at the prefix automaton one can see the effect of the infrequent trace $\sigma_5 = \langle a,c,d,e,e,f,h,i \rangle$. The detected escaping arc $\langle a,c,d,e,e,e \rangle$ is a direct result of the trace σ_5, i.e., an escaping arc that would not exist if we consider the log $L' = [\sigma_1^{1435}, \sigma_2^{946}, \sigma_3^{764}, \sigma_4^{54}]$. On the other hand, σ_5 contains the prefix $\langle a,c,d,e,e \rangle$ that otherwise, considering L', would be a escaping arc on a highly weighted state $\langle a,c,d,e \rangle$ (with its consequent effect on the metric value). A robust precision checking approach should be affected as little as possible at this kind of situations.

In the literature, several approaches have been proposed to detect noisy and anomalous traces within event logs, e.g., [24, 25]. This is not the goal of the approach presented in this section, but to extend the escaping arc detection theory in order to incorporate the weight component, and to do that in an arc-level way. Therefore, the filtering of anomalous traces from the event log can be considered a pre-processing step prior to the precision checking.

In order to increase the robustness of the approach presented in Chapter 5, in this section we propose the use of a *cut threshold* to remove states of the prefix automaton based on their weight, as a preliminary step to the precision computation. In other words, states in the log having a weight under that threshold are considered out of the observed behavior for the precision checking purposes. For example, if we consider the state $\langle a,c,d,e \rangle$ in Figure 7.2, we see that 99.9% of its behavior follows

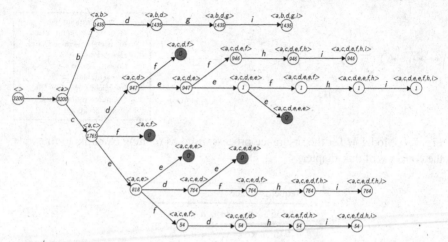

Fig. 7.2: Prefix automaton for the event log L and model M as it is presented in Chapter 5.

the f activity, being e a promising candidate to be cut. The cut threshold proposed in this section is defined to be parametric at arc-level:

- *Parametric*: The cut threshold is based on a *cut factor* $r \in [0,1]$. The cut factor is established according to the level of robustness desired. For instance, a high cut factor is used to compute the precision using only the main behavior of the system, reaching a higher level of abstraction. On the other hand, a low cut factor cuts only the most extreme cases. When the cut factor r is 0 there is no cut states, and all the behavior is considered for the precision checking.
- *Arc-Level*: The cut threshold is applied at arc-level, and not at trace level. In other words, to decide if at the state σ_t with the incoming arc $q = (\sigma_s, a, \sigma_t)$ must be cut we consider only the context of that arc, i.e., the weight of target state σ_t and the weight of its predecessor σ_s. The cut threshold is defined as the weight of the source state multiplied by the cut factor, i.e., $\omega(\sigma_s) \cdot r$. A state is cut if its weight is less or equal than the cut threshold, i.e., $\omega(\sigma_t) \leq \omega(\sigma_s) \cdot r$. Notice that, the cut threshold is defined locally to the arc, and not globally. Therefore, the cut does not depends on the weight of the overall automaton but only on the weights of the part of the process the arc refers to.

When a state $\sigma \in S_{LM}$ is defined as cut state, all the states in S_{LM} descendants of σ are also considered cut.

Figure 7.3 shows the prefix automaton of L and M when we apply a cut factor of $r = 0.03$. The automaton shows some differences with the original automaton in Figure 7.2, where the cut states are represented in a different color than the escaping states. For example, the state $\langle a,c,d,e,e \rangle$ is a cut state because its weight do not overpass the threshold $\omega(\langle a,c,d,e,e \rangle) \leq \omega(\langle a,c,d,e \rangle) \cdot r$, i.e., $1 \leq 947 \cdot 0.03 = 28.41$. Consequently, all the states after $\langle a,c,d,e,e \rangle$ are also cut states, i.e., $\langle a,c,d,e,e,f \rangle$, $\langle a,c,d,e,e,f,h \rangle$ and $\langle a,c,d,e,e,f,h,i \rangle$.

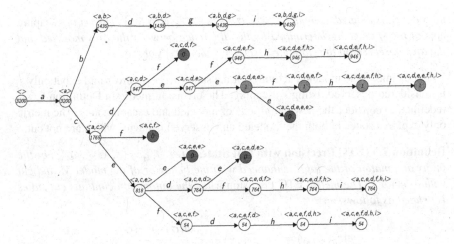

Fig. 7.3: Prefix automaton for the event log M and model M considering a cut factor $r = 0.03$.

Formally, cut states are defined as follows:

Definition 7.1 (Cut States) *Let $\mathscr{A}_{LM} = (S, A, \nearrow, \omega, \langle \rangle)$ be the prefix automaton of the log L enhanced with the behavior of the model M. Let $(\sigma_s, a, \sigma_t) \in \nearrow$ be an arc where $\sigma_t \in S_{LM}$ is a state of both log and model. Given a cut factor r, the prefix σ_t is a cut state if :*

- *its weight is less or equal than the cut threshold for that state, i.e., $\omega(\sigma_t) \leq \omega(\sigma_s) \cdot r$.*
- *any of the prefixes of σ_t is already a cut state, i.e., $\exists \sigma_a \cdot \sigma_b = \sigma_t : \sigma_a$ is a cut state.*

The set of cut states is represented as S_r. The prefix automaton of the log L enhanced with the behavior of the model M and cut with a threshold r is denoted as $\mathscr{A}_{LM}^r$.

Notice that, by definition the empty sequence $\langle \rangle$ is never a cut state because it has no incoming arc. Notice also that, when the threshold factor r is 0, no state in the log is cut and the approach behaves as the one presented in Chapter 5.

As it has been mentioned before, the cut states are considered to be out of the log behavior. Therefore, although they appear on the log, they are counted as escaping states for the precision computation. The definitions of *modeled*, *observed* and *escaping* in Definition 5.4 and Definition 5.5 are adapted to reflect this new scenario.

Definition 7.2 (Escaping Arcs with Cut States) *Let $\mathscr{A}_{LM}^r = (S, A, \nearrow, \omega, \langle \rangle)$ be the prefix automaton of the log L enhanced with the behavior of the model M and cut with a factor r. Let the prefix $\sigma_s \in (S_{LM} \setminus S_r)$ represent a state of both log and model and not cut by r. The set of arcs observed in σ_s represent those activities executed and consequently recorded on the log L but not cut by r, i.e., $obs(\sigma_s) = \{(\sigma_s, a, \sigma_t) \in \nearrow | \sigma_t \in (S_{LM} \setminus S_r)\}$. The set of activities modeled by the model M in* *the state σ_s is defined as $mod(\sigma_s) = \{(\sigma_s, a, \sigma_t) \in \nearrow \mid \sigma_t \in S_M\}$. The set of escaping arcs of the state σ_s is determined by the difference between the arcs modeled and the arcs observed on the state, i.e., $esc(\sigma_s) = mod(\sigma_s) \setminus obs(\sigma_s)$.*

For example, the modeled activities of the state $\langle a, c, d, e \rangle$ are f and e, but only f is considered observed, being e escaping. The precision metric in Definition 5.6 is redefined to consider the new definitions of modeled and escaping arcs. The metric only explores states in both the modeled and observed behavior which are not cut.

Definition 7.3 (ETC Precision with Cut States) *Let $\mathscr{A}_{LM}^{r} = (S, A, \nearrow, \omega, \langle\rangle)$ be the prefix automaton of the log L enhanced with the behavior of the model M and cut with a factor r. The metric ETC Precision when the automaton contains cut states is defined as follows:*

$$etc_p(\mathscr{A}_{LM}^{r}) = 1 - \frac{\sum_{\sigma \in (S_{LM} \setminus S_r)} \omega(\sigma) \cdot |esc(\sigma)|}{\sum_{\sigma \in (S_{LM} \setminus S_r)} \omega(\sigma) \cdot |mod(\sigma)|}$$

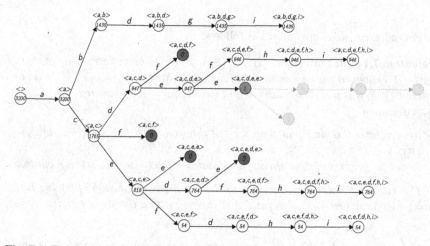

Fig. 7.4: Escaping arcs of the prefix automaton for the event log L and model M, cut with a factor $r = 0.03$.

Figure 7.4 illustrates the states and escaping arcs considered in the metric for the running example L and M. The five escaping states considered are represented in color: the four states in the same color indicate states in the model but not in the log, and the state in different color represents a cut state. The blurry states indicate the states not explored by the precision metric due to being cut states.

The metric calculation for automaton $\mathscr{A}_{LM}^{0.03}$ between L and M with a cut factor of 0.03 is:

$$etc_p(\mathscr{A}_{LM}^{0.03}) = 1-$$

$$\frac{3200 \cdot 0 + 3200 \cdot 0 + 1435 \cdot 0 + 1435 \cdot 0 + 1435 \cdot 0 + 1435 \cdot 0 + 1765 \cdot 1 + \ldots}{3200 \cdot 1 + 3200 \cdot 2 + 1435 \cdot 1 + 1435 \cdot 1 + 1435 \cdot 1 + 1435 \cdot 0 + 1765 \cdot 3 + \ldots}$$

$$\frac{\ldots + 947 \cdot 1 + 947 \cdot 1 + 946 \cdot 0 + 946 \cdot 0 + 946 \cdot 0 + 818 \cdot 1 + 764 \cdot 1 + \ldots}{\ldots + 947 \cdot 2 + 947 \cdot 2 + 946 \cdot 1 + 946 \cdot 1 + 946 \cdot 1 + 818 \cdot 3 + 764 \cdot 2 + \ldots}$$

$$\frac{\ldots + 764 \cdot 0 + 764 \cdot 0 + 764 \cdot 0 + 54 \cdot 0 + 54 \cdot 0 + 54 \cdot 0 + 54 \cdot 0}{\ldots + 764 \cdot 1 + 764 \cdot 1 + 764 \cdot 0 + 54 \cdot 1 + 54 \cdot 1 + 54 \cdot 1 + 54 \cdot 0}$$

$$= 1 - \frac{5241}{31498} = 1 - 0.17 = 0.83$$

The metric result for the automaton $\mathscr{A}_{LM}^{0.03}$ is 0.83, and as expected, it differs from the results of the automaton $\mathscr{A}_{LM}$ with no cut where $etc_p(\mathscr{A}_{LM})$ is 0.86. The difference is explained because $\mathscr{A}_{LM}^{0.03}$ contains an escaping state ($\langle a,c,d,e,e \rangle$) is a state with high weight ($\langle a,c,d,e \rangle$ with weight 947), a situation that does not occur in $\mathscr{A}_{LM}$, causing the metric to decrease.

7.3 Confidence on Precision

Given an event log and a model, the metric ETC Precision (etc_p) estimates the degree of precision of the system. However, together with a metric, sometimes it is convenient to provide also a *confidence* value, indicating the possible variability of the metric in the future. In this section we provide this confidence value in terms of a *confidence interval*, i.e., an *upper confidence value ($etc_p^\top$)* and a *lower confidence value ($etc_p^\perp$)* estimating variability over the computed metric. A narrow interval indicates that the metric should not vary significantly. On the other hand, a wide interval reflects the opposite: a low confidence in the metric provided, whose value could change drastically in the future.

Both confidence values presented in this section are defined in terms of k, a parameter representing the future behavior to be considered. A low k is used to compute the confidence in a near future, whereas with a large k, a larger incoming behavior is considered, and thus a longer term future is contemplated. We will implement k as number of traces to consider.

Notice that, both confidence values presented in this section are approximated, i.e., they do not provide real bounds over the metric, but instead aim at estimating them with simple and intuitive heuristics which can be computed in a systematic manner.

7.3.1 Upper Confidence Value

Given an automaton $\mathscr{A}_{LM}^r$ and the parameter k representing the future behavior to consider, the upper value of the confidence interval is computed considering a best possible scenario. That is, all the future behavior aims only to *cover* existing escaping arcs. In other words, each j where $1 \leq j \leq k$ represents a new occurrence in an escaping arc, increasing the weight of the both states of the arc. In that sense, each j can be seen as a future trace reaching an escaping arc $q = (\sigma_s, a, \sigma_t)$. Both if the target state σ_t is not in the log or it has been cut, a trace j represents a new occurrence of it in the log. If the number of occurrences is enough to overpass the cut threshold (cf. Section 7.2), the arc is no longer considered escaping.

The cost in number of occurrences needed to change a escaping arc $q = (\sigma_s, a, \sigma_t)$ from escaping to non escaping depends on the cut factor r considered, i.e., $\omega(\sigma_t) \leq \omega(\sigma_s) \cdot r$. Notice that, states in the model but not in the log (i.e., with weight 0), are under that threshold, no matter what r is considered. The technique presented in this section estimates the gain (i.e., the precision increase) of covering each escaping arc, and maximizes the total gain considering k as the maximum number of occurrences used to cover escaping arcs. For that, we first define the *cost* and the *gain* of covering an escaping arc.

Definition 7.4 (Cost and Gain of Covering and imprecision) *Let $q = (\sigma_s, a, \sigma_t)$ be an escaping arc of the automaton $\mathscr{A}_{LM}^r$. The cost of covering q, denoted as $cost(q) = c$ with $c \in \mathbb{N}$, is the minimum c that satisfies $\omega(\sigma_t) + c > (\omega(\sigma_s) + c) \cdot r$, i.e., it overpasses the cut threshold. The gain of covering the escaping arc q is defined as $gain(q) = \omega(\sigma_s)$, i.e., the gain of reducing in one the number of escaping arcs of the source state σ_s.*

By inspecting the formula of the ETC Precision metric (cf. Definition 7.3), one can see why the gain of covering the escaping q is defined as $\omega(\sigma_s)$: if the state σ_s has one less escaping arc, the number of escaping arcs become $|esc(\sigma_s)| - 1$. Since this number is multiplied by $\omega(\sigma_s)$ in the numerator part of the equation, the numerator will be reduced exactly in $\omega(\sigma_s)$.

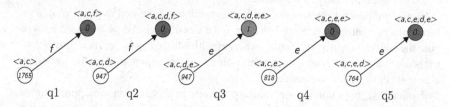

Fig. 7.5: Close up of the escaping arcs of the automaton $\mathscr{A}_{LM}^{0.03}$ for the running example M and L.

Let us consider the escaping arcs of the running example $\mathscr{A}_{LM}^{0.03}$, enumerated in Figure 7.5. The cost of the escaping arc q_2 is 30, i.e., 30 traces need to reach that arc to overpass the threshold. Instead, 29 are not enough because $\omega(\langle a,c,d,f \rangle) + 29 \not> (\omega(\langle a,c,d \rangle) + 29) \cdot 0.03$. On the other hand, the escaping arc q_3 only needs 29. That contrasts with the the cost of the escaping arc q_1 – that needs 55 – because of the weight of $\langle a,c \rangle$ – that is 1765. The gain of covering q_2 and q_3 is the same, i.e., 947. On the other hand, the gain of covering q_1 is 1765.

Once the gain and cost of covering an escaping arc are defined, the maximum gain obtained with k occurrences can be formulated. This problem is analogous to the well known *Knapsack* problem [39], which can be solved using *binary integer programming (BIP)* [81]. The following BIP model encodes the maximum gain obtained covering imprecisions with at most k occurrences:

1. *Variables:* The variable X_i denotes if the imprecision i is covered or not.

$$X_i \in \{0,1\}$$

2. *Constraints:* The total cost cannot exceed the number of occurrences.

$$\sum_{i \in esc(\mathscr{A}_{LM}^r)} cost(i) \cdot X_i \leq k$$

3. *Cost function:* Maximize the gain.

$$\max \sum_{i \in esc(\mathscr{A}_{LM}^r)} gain(i) \cdot X_i$$

Once the optimization problem is formulated, the upper confidence value can be defined as follows:

Definition 7.5 (Upper Confidence Value) *Let $\mathscr{A}_{LM}^r = (S, A, \nearrow, \omega, \langle \rangle)$ be the prefix automaton of the log L enhanced with the behavior of the model M and cut with a factor r. Let k be the future parameter to consider. Let numerator and denominator be the numerator and denominator of the metric ETC Precision in Definition 7.3, i.e., $etc_p(\mathscr{A}_{LM}^r) = 1 - \frac{numerator}{denominator}$. Let $gain_{max}$ be the result obtained using the optimization problem modeled above. The* upper confidence value *is defined as:*

$$etc_p^\top(\mathscr{A}_{LM}^r, k) = 1 - \frac{numererator - gain_{max}}{denominator}$$

Following with the running example of $\mathscr{A}_{LM}^{0.03}$ and considering a future parameter $k = 24$, the only escaping arc with cost lower enough to be covered with this k is q_5. The gain of covering this escaping arc is 764. This value is subtracted from *numerator*, providing an upper interval value of 0.85 for this scenario.

7.3.2 Lower Confidence Value

The idea for computing the lower confidence value is similar to the upper value. However, in this case the k representing the future does not cover escaping arcs, but produce the rising of new escaping arcs instead. In other words, we consider k new observed traces. Each one of those k traces introduces m new observed states. And each one of those m states causes n new escaping arcs to appear. The lower confidence value is the decrease on the precision metric caused by the escaping arcs introduced by these new k traces.

The number of escaping arcs introduced is directly related with the number of observed states introduced by the new traces, i.e., m. Longer traces cause more escaping arcs, and as a result a lower confidence value. The are several alternatives of the value of m, e.g., the length of the longest trace in the log, or just an arbitrary number. However, for statistical consistency, the m considered in this approach corresponds with the average length of the traces in the log. Moreover, the number of new escaping arcs introduced for each state (i.e., n) is also directly related with the confidence value obtained. Given that we are considering a worst case scenario, we consider n to be $|A - 1|$, i.e., all activities are escaping arcs except the one followed by the trace. Given all these considerations, the lower confidence value is defined as follows:

Definition 7.6 (Lower Confidence Value) *Let $\mathscr{A}_{LM}^r = (S, A, \nearrow, \omega, \langle \rangle)$ be the prefix automaton of the log L enhanced with the behavior of the model M and cut with a factor r. Let numerator and denominator be the numerator and denominator of the metric ETC Precision in Definition 7.3, i.e., $etc_p(\mathscr{A}_{LM}^r) = 1 - \frac{numerator}{denominator}$. Let k be the future to consider, and let avg be the average length of the traces in the log L. The* lower confidence value *is defined as:*

$$etc_p^{\perp}(\mathscr{A}_{LM}^r, k) = 1 - \frac{numerator + (k \cdot avg \cdot |A - 1|)}{denominator + (k \cdot avg \cdot |A|)}$$

For instance, following with the running example, being $avg = 6$ the average length, $A = 8$ the number of activities, and considering $r = 0.03$ and $k = 24$, the lower bound in this case is:

$$etc_p^{\perp}(\mathscr{A}_{LM}^r, 24) = 1 - \frac{5241 + (24 \cdot 6 \cdot 7)}{31498 + (24 \cdot 6 \cdot 8)} = 0.81$$

7.4 Experimental Results

This section illustrate experimentally some of the concepts presented on this chapter. Table 7.2 contains the results of applying the proposed approach to a set of large datasets. For comparative reasons, the datasets are some of the ones used in Chapter 6, obtained from the same public repository. As it is mentioned in Chapter 6, these datasets cannot be handled by precision approaches such as $a'b$ [77]. The experimental setting is based on variations of the $a32f0n00\_5$ and $t32f0n00\_5$ datasets, and the experiments focus on illustrating how the growth of a log influences the metric and its confidence, given a particular selection of the stability and confidence parameters presented in this chapter. The column with pX reports the percentage of the log considered in each case, i.e. $p100$ represents the original $a32f0n00\_5$ log, while logs pX with $X < 100$ correspond to slices of the original log, e.g., $p20$ contains the first 20% of the original log traces. Logs pX with $X > 100$ are obtained by choosing with uniform distribution among the existing traces in the log the extra traces needed to achieve the desired size.

Bench		|Log|	r	k	etc_p	Confidence	time(s)
a32	p20	180			.543	.246 - .553 (.307)	1 / 3 / 5
	p40	360			.564	.345 - .570 (.225)	1 / 5 / 6
	p60	540			.576	.403 - .582 (.179)	1 / 7 / 11
	p80	720			.583	.441 - .587 (.146)	1 / 12 / 17
	p100	900	.05	20	.592	.470 - .595 (.125)	1 / 15 / 24
	p150	1350			.591	.504 - .595 (.091)	2 / 16 / 23
	p200	1800			.591	.523 - .595 (.072)	2 / 17 / 23
	p250	2250			.590	.534 - .594 (.060)	2 / 16 / 24
	p300	2700			.591	.544 - .594 (.050)	2 / 16 / 24
t32	p20	360			.385	.250 - .387 (.137)	2 / 67 / 121
	p40	720			.391	.305 - .392 (.087)	4 / 180 / 229
	p60	1080			.392	.330 - .393 (.063)	5 / 295 / 339
	p80	1440			.393	.345 - .394 (.049)	6 / 336 / 496
	p100	1800	.05	20	.393	.353 - .394 (.041)	6 / 390 / 550
	p150	2700			.393	.365 - .393 (.028)	6 / 411 / 562
	p200	3600			.393	.371 - .393 (.022)	7 / 429 / 572
	p250	4500			.393	.376 - .393 (.017)	9 / 440 / 579
	p300	5400			.393	.379 - .393 (.014)	9 / 443 / 581

Table 7.2: Precision metric value, confidence and computation time for incremental benchmarks.

The models used are the ones obtained from discovering a Petri net through the ILPMiner [93]. The wide spectrum of the set of benchmarks presented makes it possible to illustrate the evolution of the approach presented in this chapter and can be considered as a real situation in an information system where trace sets are evaluated on a regular basis, e.g., monthly.

A first conclusion on the table is the stability of the approach with respect to the size of the log. Notice that the etc_p value tends to increase as new behavior is considered, e.g., between *p20* and *p100* there is a difference of 0.05. However, this difference is extremely small considering that between *p20* and *p100* there is a 500% increment in the observed behavior. In addition, the more traces are included in the previously observed behavior, the closer the metric value is to stabilizing. The second conclusion to extract from this table is the dependency between the traces considered and the confidence in the metric, i.e., increasing the size of the trace set considered results in a narrower confidence interval.

The next set of experiments are designed to illustrate the influence of confidence parameter k and the cutting parameter r in the proposed approach.

In chart 7.6a, three process models are considered: *ext_c1_01*, *ext_c1_02* and *ext_c1_03*. These benchmarks have been created using the PLG tool [37]. This tool allows to create configurable and generic benchmarks, containing all the common patters appearing in any workflow model, e.g., choice, parallelism, sequence, etc. For the experiment, each one of the logs considered contains 15000 traces. Bench-

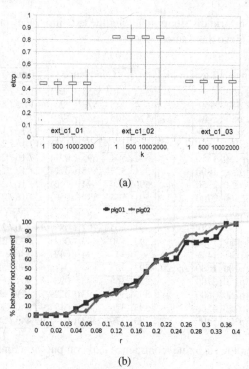

(a)

(b)

Fig. 7.6: Charts showing the effects of different parameters for the confidence value.

marks 01 and 03 denote standard processes, with great difference between model and log behavior (and thus exhibiting low precision). On the other hand, process model *ext_c1_02* is a simpler model which describes accurately the behavior reflected in the log, i.e., the precision value is high. The chart illustrates the influence in the approach when considering diverse future windows, i.e., four different k values: $1, 500, 1000$ or 2000 new traces to consider. As it is reflected in the experiments, the future considered has no influence on the metric value, but it is relevant on the confidence value over the metric. The possibility of variation for the metric considering a really near future (i.e. $k = 1$) is practically zero. However, when considering farther futures, this possibility increases, e.g., considering a k value of 2000 (approx. 15% of the log) the confidence in the metric is substantially low. Notice that, as expected, the confidence interval is not symmetric.

Chart 7.6b illustrates the relation between the cut factor r and the percentage of the process behavior considered to compute the precision. Two generic and representative process, *plg_01* and *plg_02*, have been created using PLG tool, and different values of r have been tested. The conclusion we obtained is that for these processes, lower values of r (i.e., less than 0.04) can be used to *polish* the effects produced by noisy traces, while greater values of r (not considering more than 10% of the process behavior for computing the precision) should be used if the emphasis is in computing precision on the most frequent parts of the process. Values greater than 0.4 does not make any sense, due to the 100% of process is discarded.

7.5 Summary

Given an event log and a process model, the approach to measure precision based on escaping arcs presented in Chapter 5 is sensitive to the presence of infrequent behavior. This chapter presented an extension of the escaping arc theory to increase its robustness, by discarding exceptional behavior from the precision computation. Furthermore, the chapter presented a confidence interval over the precision metric, indicating the likelihood of the metric to drastically change when more observed behavior is considered.

Chapter 8
Assessing Severity

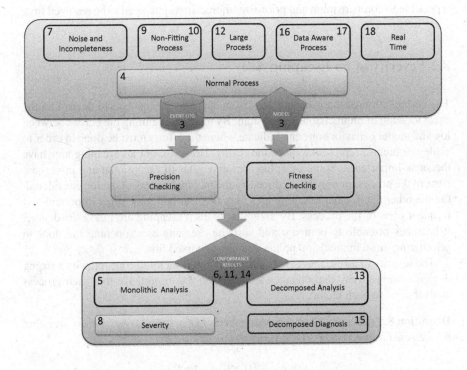

The previous chapters presented a technique to detect precision issues between process models and logs. However, not all the detected precision problems have the same severity. A good diagnosis tool must evaluate and categorize them according to their importance. This chapter provides an assessment approach to measure the severity of the escaping arcs, quantifying their severity using four dimensions:

© Springer International Publishing AG 2016
J. Munoz-Gama: Conf. Check. ... in Process Mining, LNBIP 270, pp. 75–84, 2016.
DOI: 10.1007/978-3-319-49451-7_8

weight, alternation, stability, and criticality. In later chapters, we will extend the precision detection to handle non-fitting scenarios.

8.1 Introduction

The precision based on escaping arcs aims to estimate the effort needed to achieve a perfectly precise system, i.e., resolving all the escaping arcs result in a model that precisely describes the observed behavior. However, analyzing the cause of an escaping arc and eventually fixing it requires time and it is a resource-consuming task. One only wants to invest time and effort in resolving deviations which represent a clear and severe precision problem. On the other hand, it may be considered a waste of resources to fix escaping arcs that are likely to disappear in a near future. In this chapter we propose a multi-factor severity assessment of the escaping arcs, making it possible to compare them and prioritize imprecisions that need to be resolved first.

8.2 Severity of an Escaping Arc

The computation of the escaping arcs is an accurate mechanism to determine the exact location of imprecisions of a system. By means of defining the border between log and model behaviors one can indicate where the efforts must be done in order to achieve a precise representation of the reality. However, not all escaping arcs have the same importance. Some escaping arcs may refer to exceptional and infrequent parts of the process or may be produced by the incompleteness of the log considered. On the other hand, other escaping arcs may be clear and may affect important and frequent parts of the process. By assigning to each escaping arc a *severity* degree, it becomes possible to compare and sort the escaping arcs, opening the door to prioritizing those imprecision points that must be fixed first.

The severity of an escaping arc is a complex multi-factored concept, with a strong degree of subjectivity that changes according to the importance that each process analyst gives to each factor.

Definition 8.1 (Severity of an Escaping Arc) *Let $q \in esc(\mathscr{A}_{LM}^r)$ be an escaping arc detected in the system. The* severity *of q is defined as:*

$$sev(q) = f(fact_1^q, \ldots, fact_n^q)$$

where $fact_1^q, \ldots, fact_n^q$ correspond to the factor values for the escaping arc q, and f is a user-defined function that weights the n factors.

In this section, we define a standard weighting function as the arithmetic mean of all the factor values considered, giving the same importance to all the factors.

Alternative weighting functions can be defined assigning higher weights to some factors, and lower weights (or even 0) to others.

The definition and selection of the different factors considered in the severity function are also open to subjectivity, and may vary from one context to another. In this book we propose a four-factored severity with *weight*, *alternation*, *stability* and *criticality* being the four factors considered.

8.2.1 Weight of an Escaping Arc

The first factor proposed to assess the severity of an escaping arc is the weight. Escaping arcs occurring on highly weighted states denote problems in important parts of the process. These escaping arcs occur frequently during the process executions, and should be in the top of the list of problems to fix. On the other hand, escaping arcs that appear on states with low weight indicate problems in infrequent parts of the process, with low representation in the process executions.

The *weight factor* considers the weight of the state where the escaping arc occurs, normalizing it with the maximum weight of all the states in the automaton. Therefore, the factor value ranges from 0 to 1.

Definition 8.2 (Weight Factor) *Let $q = (\sigma_s, a, \sigma_t) \in esc(\mathscr{A}_{LM}^r)$ be an escaping arc detected in the automaton $\mathscr{A}_{LM}^r = (S, A, \nearrow, \omega, \sigma_0)$. Let max be the maximum weight of all the states in S, i.e., $\forall \sigma \in S, max \geq \omega(\sigma)$. The weight factor of the escaping arc q is defined as:*

$$fact_w^q = \frac{w(\sigma_s)}{max}$$

For example, let us consider the escaping arc q_1 of the running example $\mathscr{A}_{LM}^{0.03}$. The maximum weight of all the states of $\mathscr{A}_{LM}^{0.03}$ corresponds with the state $\langle\rangle$, i.e., $max = 3200$. Therefore, the weight factor for q_1 is $fact_w^{q_1} = \frac{w(\langle a,c\rangle)}{max} = \frac{1765}{3200} = 0.55$. The mid-value weight factor for q_1 contrasts with the $\frac{764}{3200} = 0.24$ value for the imprecision q_5.

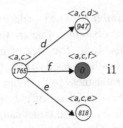

Fig. 8.1: Close-up of the escaping arc q_1 of $\mathscr{A}_{LM}^{0.03}$ and its context.

8.2.2 Alternation of an Escaping Arc

The *alternation factor* addresses situations where the system enables a set of alternatives, but only a few part of them can be considered as valid alternatives. In this case, the set of alternatives are the set of arcs in a state, while a wrong alternative is to chose to continue through an escaping arc. Situations where it is more likely to choose wrongly must have higher priority than situations where choosing an escaping arc is not so probable.

Definition 8.3 (Alternation Factor) *Let $q = (\sigma_s, a, \sigma_t) \in esc(\mathscr{A}^r_{LM})$ be an escaping arc detected in the automaton $\mathscr{A}^r_{LM} = (S, A, \nearrow, \omega, \sigma_0)$. Given a state $\sigma \in S$, let $P_E(\sigma)$ be the probability of selecting a escaping arc being in the state σ. The alternation factor of the imprecision q is defined as:*

$$fact^q_a = P_E(\sigma_s)$$

The distribution of $P_E(\sigma_s)$ depends on the assumptions taken. In the general case where no particular assumption are made, a uniform distribution is considered, i.e., given σ_s, all the outgoing arcs of σ_s have the same probability of being reached. Considering a uniform distribution the alternation factor of the escaping arc q is reformulated as:

$$fact^q_a = \frac{|esc(\sigma)|}{|mod(\sigma)|}$$

Notice that the alternation is a factor based on the source state σ_s of the escaping arc, not the target escaping state σ_t. Therefore, all escaping arcs with the same source state have the same alternation value. The alternation factor measures the amount of alternation in each escaping arc. For instance, the alternation value for the escaping arc q_1 is $fact^{q_1}_a = \frac{1}{3} = 0.33$, denoting a mid-low probability of selecting a escaping state – only $\langle a, c, f \rangle$ in this case– from the state $\langle a, c \rangle$.

8.2.3 Stability of an Escaping Arc

The third factor proposed – the *stability factor* – addresses the stability or equilibrium of an escaping arc, i.e., the probability of an arc to stop being an escaping arc after applying a little *perturbation* to it. The idea of introducing perturbations in order to estimate some property has been used successfully in other fields, such as the measurement of community robustness [49]. In our setting, a perturbation over an escaping arc is represented by a small set of traces going across the escaping arc, modifying the weight of both source and target state, and possibly changing its status from escaping to normal arc. The number of traces used as perturbation – and represented by z – are defined proportionally to the weight of the source state by means of the *perturbation intensity* τ.

Definition 8.4 (Perturbation of an Escaping Arc) *Let $q = (\sigma_s, a, \sigma_t) \in esc(\mathscr{A}_{LM}^r)$ be an escaping arc detected in the automaton $\mathscr{A}_{LM}^r = (S, A, \nearrow, \omega, \sigma_0)$. Let $\tau \in [0, 1]$ be the perturbation intensity considered. The number of traces considered as perturbation is defined as $z = \lceil \omega(\sigma_s) \cdot \tau \rceil$. Let $l \in \mathbb{N}$ be the smallest number such that the equation $\omega(\sigma_t) + l > (\omega(\sigma) + z) \cdot r$ is satisfied, i.e., l defines the minimum number of traces the arc q must receive in order to overpass the cut threshold r after considering z traces, and change from escaping to normal arc.*

For example, let us consider the escaping arc q_1 of $\mathscr{A}_{LM}^{0.03}$ where $r = 0.03$. Considering a perturbation intensity of $\tau = 0.06$, the perturbation traces are $\lceil 1765 \cdot 0.06 \rceil = 106$, and consequently, l is $\lceil ((1765 + 106) \cdot 0.03) - 0 \rceil = 57$.

The stability factor is defined as the probability of an escaping arc to remain escaping after perturbing it.

Definition 8.5 (Stability Factor) *Let $q = (\sigma_s, a, \sigma_t) \in esc(\mathscr{A}_{LM}^r)$ be an escaping arc of $\mathscr{A}_{LM}^r = (S, A, \nearrow, \omega, \sigma_0)$. Let τ be the perturbation intensity, and let z and l be the corresponding perturbation traces and minimum, respectively. The stability factor of the escaping arc q is the probability of q to remain escaping after considering z new traces, i.e.,*

$$fact_{s\tau}^q = P_q^z(< l) = \sum_{i=0}^{l-1} P_q^z(= j)$$

where $P_q^z(< x)$ and $P_q^z(= x)$ represent the probability that the arc q receives less than x (or exactly x) of the new z traces considered in this point.

Let p_q define the probability that a new trace crosses the escaping arc q. Let $1 - p_q$ be the probability that the trace follows one of the other successor states of σ_s. According to the binomial distribution [27], the stability factor can be expressed as:

$$fact_{s\tau}^q = \sum_{i=0}^{l-1} \binom{z}{i} (p_q)^i (1 - p_q)^{z-i}$$

The formula can be understood as follows: in order to q to remain escaping i successes $(p_q)^i$ and $z - i$ failures $(1 - p_q)^{z-i}$ are needed. However, the i successes can occur anywhere among the z traces, and there are $\binom{z}{i}$ different ways of distributing i successes in a sequence of z traces.

The probability p_q may depend on the assumptions taken. Again, if no knowledge regarding the distribution of the log is assumed, a uniform distribution is considered. Therefore, if c is the number of successors states of σ_s, the probability of each successor state is $1/c$, and the formula can be rewritten as:

$$fact_{s\tau}^q = \sum_{i=0}^{l-1} \binom{z}{i} \left(\frac{1}{c}\right)^i \left(1 - \frac{1}{c}\right)^{z-i}$$

In the running example $\mathscr{A}_{LM}^{0.03}$, given the escaping q_5 and considering $r = 0.03$ and $\tau = 0.06$, the stability factor results in 0.67, being $z = 46$ and $l = 25$. This factor reflects that this escaping arc has a mid-probability of disappearing in the close future. This contrasts with the stability 1 obtained from the escaping arc q_1, with the same r and τ parameters, reflecting a really stable imprecision.

8.2.4 Criticality of an Escaping Arc

Finally, the *criticality factor* introduces domain knowledge into the severity assessment. Different activities constitute different levels of criticality within the process. For example, the possible consequences of *CheckDateFormat* action are not the same as the *TransferMoney* action, so neither are their criticality. Escaping arcs in the model allowing to execute *TransferMoney* – an action never observed in the log on that state – must have a higher severity and should be analyzed before than escaping arcs allowing *CheckDateFormat*.

The criticality factor relies on a function *crit* defined externally by a domain expert, assessing the criticality of the activities of the process. If this function is not defined, we consider a function where all activities have the same criticality. The criticality function is inspired by *cost function* used by the alignment conformance algorithms to weight the different types of misalignments [18, 17].

Definition 8.6 (Criticality Factor) *Let* $q = (\sigma_s, a, \sigma_t) \in esc(\mathscr{A}^r_{LM})$ *be an escaping arc detected in the automaton* $\mathscr{A}^r_{LM} = (S, A, \nearrow, \omega, \sigma_0)$. *Let* $cri : A \to [0,1]$ *be the function that assess the criticality of the process activities. The* criticality factor *of the escaping arc q is defined as:*

$$fact^q_c = crit(a)$$

In our running example, we consider that *Check Liability (f)* is a relative cheap operation where the database is queried automatically, whereas *Consult Expert (e)* is a much costly operation that involve a human consultation. Consequently, considering $crit(e) = 0.9$ and $crit(f) = 0.2$, the escaping arcs q_3, q_4, q_5 are considerably more critical than q_1, q_2.

8.2.5 Visualizing the Severity

The severity of an escaping arc is the result of combining the different factors through the function f, that assign weights to each factor (cf. Definition 8.1). For example, the standard severity proposed in this chapter for an escaping arc q is defined as $sev(q) = mean(fact^q_w, fact^q_a, fact^q_{s\tau}, fact^q_c)$. This value can be used to visualize the severity escaping arcs of a system in more advanced ways than simply displaying the value next to the arc. For example, the set of escaping arcs can be partitioned according to their severity ($[0.0, 0.3) = low$, $[0.3, 0.4) = mid$, and $[0.4, 1.0) = critical$), and using an intuitive traffic light color scheme (i.e., green, yellow, red), an analyst can easily identify visually the different severity degrees.

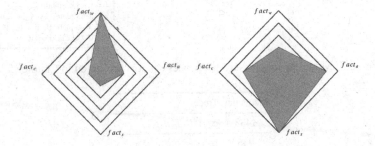

Fig. 8.2: Two escaping arcs with different factor value distribution.

However, there are situations where a factor-grained analysis of the severity may be more adequate, i.e., analyzing the factors independently, instead of aggregating them on a unique value. For those cases, displaying the factor results in a *radar chart* provides a better understanding of the different factor distribution composing the severity of an escaping arc. For example, Figure 8.2 shows two possible escaping arcs with different factor distribution. The left chart corresponds to a really frequent escaping arc. However, the situation of that escaping arc is really unstable and the possibilities of choosing badly in that situation are really few. The second escaping arc, shown in the chart on the right, is much more sever in general terms than the first one. It corresponds to a more stable and dangerous situation, but it is less frequent than the first one.

8.2.6 Addressing Precision Issues based on Severity

Finally, once the severity of all the escaping arcs has been evaluated, the results are taken into consideration in order to prioritize the future actions over the process. Let us consider the example insurance process seen in Chapter 5 (cf. Figure 8.3 and Table 8.1). In this small example, 5 escaping arcs are detected denoting 5 precision problems. However, the severity analysis determines that the $\langle a, c, f \rangle$ needs to be urgently addressed: although its alternation is lower than the others (0.33 in comparison with 0.5), its weight is considerably higher than the others. Moreover, the domain experts determine that f (Check Liability) is a highly critical activity, and its execution needs to be precisely controlled. Therefore, the model is changed to forbid the execution of f without executing e previously. On the other hand, the domain experts decided that, addressing the escaping arcs in $\langle a, c, d, e, e, e \rangle$, may be more costly than the possible benefits from the update, given the low weight of the escaping arc.

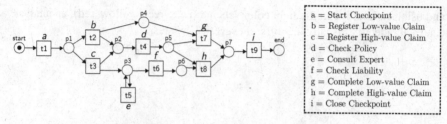

Fig. 8.3: Model M for a insurance process as it is presented in Chapter 5.

Frequency	Trace
1435	$\sigma_1 = \langle a,b,d,g,i \rangle$
946	$\sigma_2 = \langle a,c,d,e,f,h,i \rangle$
764	$\sigma_3 = \langle a,c,e,d,f,h,i \rangle$
54	$\sigma_4 = \langle a,c,e,f,d,h,i \rangle$
1	$\sigma_5 = \langle a,c,d,e,e,f,h,i \rangle$

Table 8.1: Event log L for a insurance process as it is presented in Chapter 5.

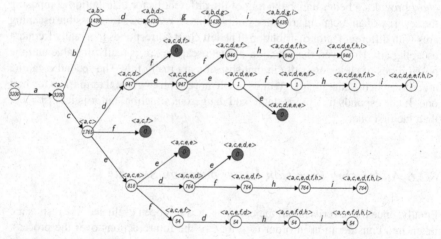

Fig. 8.4: Prefix automaton for the event log L and model M as it is presented in Chapter 5.

8.3 Experimental Results

Charts 8.5a and 8.5b are used to illustrate the the severity concept. Four generic models (sev01, sev02, sev03 and sev04) have been created using the PLG tool, which contain the most common structures in workflow models. For each model,

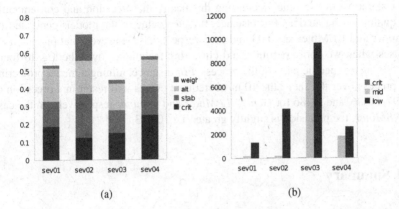

(a)   (b)

Fig. 8.5: Severity analysis of four escaping arcs, and partitioning of the detected escaping arcs in three severity categories: low, mid, and critical.

a severity analysis has been performed, where each factor has received the same weight.

The same values of r and τ have also been assigned for all the models (0.01 and 0.06, respectively). The criticality value assigned to each of the tasks is different depending on the task and the model. In chart 8.5a, we selected the most severe escaping arc of each process, and show the distribution of each one of the four factors. This chart illustrates the distribution of the weight of each factor in the final severity value, in this particular setting. In addition, it also illustrates that, given the normalization introduced in the definition of each one of the factors, it is possible to compare the severity between imprecision of different processes, e.g., given a system containing the four process models, and given the current setting, the escaping arc shown in the second bar (which corresponds to model sev02) should be tackled first. In chart 8.5b, the same four processes are considered. In this case, escaping arcs of each process are classified in three categories: *low* (less than 0.3), *mid* (between 0.3 and 0.4) and *critical* (greater than 0.4). Notice that, in this particular scenario, the number of critical escaping arcs (the ones that should be tackled urgently) is small (approx. 10 imprecisions for each process) compared to the total number of escaping arcs. Based on this, a process analyst can now focus their improvement efforts on these ca. 40 imprecisions rather than all 20000 imprecisions.

Finally, the approach presented in this chapter has been tested in a real world scenario. The scenario is taken from a Dutch Academic Hospital, and the log contains about 150.000 events in 1100 cases, where each case corresponds with a patient of the Gynecology department. The goal of the experiment is to measure the quality of the models obtained using different discovery algorithms. The process miners used in the experiment are *RBMiner*[84], *Genet*[38], *ILPMiner* [93] and *α-miner*[15]. The results illustrate that the precision of the models obtained using such miners focusing on the whole process is very low. The generated models allow almost all

the tasks most of the time, decreasing drastically the precision and consequently
the quality of the models. For instance, the etc_p value of the models generated by
α-miner and RBMiner are 0.03 and 0.01 respectively. However, that precision increases when we apply partition and clustering techniques over the log, to focus
the mining on specific parts of the process. For instance, mining a model projecting
the process over the set of the 10 most frequent events will result in a precision of
0.397, 0.386, and 0.386 for *Genet*, *ILPMiner* and α-*miner* respectively. In the case
of *RBMiner*, the precision is slightly greater, i.e., 0.423.

8.4 Summary

The chapter proposed a multi-factor measure to quantify the severity of the detected
escaping arcs, making it possible to compare them and to prioritize imprecisions that
need to be resolved first. The four factors proposed are weight, alternation, stability,
and criticality.

Chapter 9
Handling non-Fitness

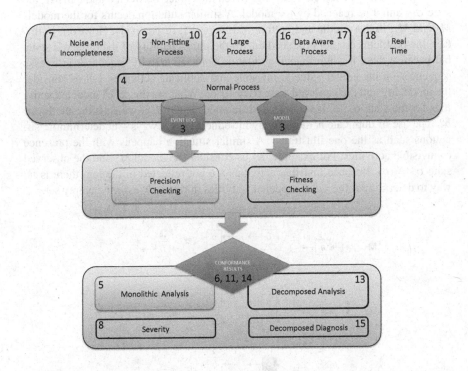

In the first part of the book, we presented a precision checking technique based on escaping arcs. However, it is not uncommon to have a certain degree of unfitness when we measure precision, e.g., invisible or duplicate activities, or simply produced by small known mismatches between model and process. This chapter provides an overview on the use of alignment techniques as a pre-processing step to

© Springer International Publishing AG 2016
J. Munoz-Gama: Conf. Check. ... in Process Mining, LNBIP 270, pp. 85–96, 2016.
DOI: 10.1007/978-3-319-49451-7_9

provide meaningful precision results in unfitting scenarios. In the next chapter, we will present some variants of this technique more suitable for some cases.

9.1 Introduction

As it has been illustrated in previous chapters, replaying observed behavior over the modeled behavior is an effective and efficient way to detect escaping arcs, and with them, detect potential imprecise points to be fixed. However, there are situations where the observed behavior cannot be replayed on the modeled behavior, i.e., the model cannot mimic the "moves" observed in the log. These problems are produced by the presence of *unfitting* traces, or *undeterministic* situations on the model. For example, let us consider the model M_1 at the top of Figure 9.1, and the state $\sigma_1 = \langle a, b, c \rangle$ observed in the log. Given the choice between b and c in M_1, the state σ_1 cannot be reached by the model. A similar situation occurs for the models M_2 and M_3, in the middle and bottom of Figure 9.1, respectively. However, in this case the problems are not produced due to the unreachability of the observed state, but because there is a non-bijective relation between activity sequences and tasks sequences on the model – due to invisible and duplicate activities. Let us consider the model M_2 and the observed state $\sigma_2 = \langle a, b \rangle$. What are the set of modeled activities for the state σ_2? c, if we consider the upper b? Or d, if we consider the lower b?. The use of duplicate activities may introduce – not always – undeterminitic situations such as the one illustrated. A similar situation happens with the presence of invisible activities. For example, let us consider the model M_3 and the observed state $\sigma_3 = \langle a \rangle$. Because invisible activities are not reflected in the log, there is no way to determine if the set of modeled activities in σ_3 are b, c, or the empty set.

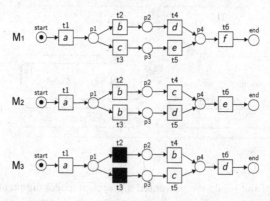

Fig. 9.1: Models to illustrate the problems with the replay produced by unfitting traces (top), the indeterminism of duplicate activities (middle), and the indeterminism of invisible activities (bottom).

The effect of problems derived from unfitting logs can be alleviated by making some assumptions over the observed behavior. For example, in Section 5.6.1 the unfitting part of the trace is considered noisy or infrequent and it is discarded for the precision metric. This drastic approach is useful when the percentage of log discarded is small, and we are not interested on providing all possible points of imprecision. However, the effect in a worst case scenario is not negligible. For example, in a log where only the first event in all traces is wrongly recorded, the whole log is discarded. Similar, heuristic assumptions could be considered to solve indeterministic situations with no formal guarantees, but still have practical applicability.

Undeterminism is produced because the escaping arcs are detected at a log level. At a model level, each task is unique, e.g., although t_2 and t_3 have the same label b, they are distinguishable in the model M_2. Therefore, a precision checking at a task level will solve indeterministic situations, transferring the responsibility to mapping correctly log traces and tasks sequences of the model. Thus, each trace in the log is translated to a complete tasks sequence in the model. For example, log traces $\langle a,b,c,e \rangle$ and $\langle a,b,d,e \rangle$ may be associated with $\langle t_1,t_2,t_4,t_6 \rangle$ and $\langle t_1,t_3,t_5,t_6 \rangle$ of model M_2, respectively.

The mapping between log traces and model traces is far from trivial. An approach may consider the use of heuristics, such as *look-ahead* [90, 28]. Informally, when the approach reaches a decision point, the algorithm looks ahead to choose the most suitable option. For example, in the trace $\langle a,b,c,e \rangle$ for the M_2, t_2 is selected because there is a c next. Similar happens when the indeterminism is introduced by invisibles, e.g., trace $\langle a,b,d \rangle$ is associated with complete tasks sequence $\langle t_1,t_2,t_4,t_6 \rangle$ in M_3. Look-ahead heuristics can be also used to associate unfitting observed states to reachable states of the model. For example, given the observed trace $\langle a,b,c,d,f \rangle$ and the model M_1, the algorithm could consider c an event wrongly recorded in the log.

Look-ahead and other heuristics are heuristics after all, and therefore, they lack formal guarantees. The decision taken heuristically may not be the optimal. Even, when the number of events considered in the look-ahead is increased, the decision may still not be optimal, e.g., the optimal path may require the reconsideration of previous decisions [17].

In this chapter we introduce a precision checking approach based on aligning observed and modeled behavior. The alignment is done at a model level, for each one of the traces in the log. The alignment techniques provide global optimal results and therefore, there are guarantees on the escaping arcs detected. Notice that, the computation cost of aligning observed and modeled behavior in a global optimal may be considerable in some cases. Therefore, there are situations where other alternatives need to be considered, for example, decomposed aligning for conformance diagnosis (cf. Chapter 12), or a heuristic replay-based approach in real-time scenarios where the time is a crucial element (cf. Chapter 18).

9.2 Cost-Optimal Alignment

The use of alignment techniques in conformance checking was first proposed by Adriansyah, van Dongen, and van der Aalst [17]. An alignment between an event log and a process model relates occurrences of activities in the log to tasks of the model. As the execution of a case is often independent from the execution of another case, the alignment is performed per traces. This is a common assumption taken in process mining techniques, and reduces the complexity of the analysis.

For each trace in an event log that fits a process model, each *move* in the trace (i.e., an activity observed in the log) can be mimicked by a *move* in the model (i.e., a task executed in the model). However, this is not the case if the trace does not fit the model perfectly. We use the symbol $\gg$ to denote "no move" in either log or model.

Definition 9.1 (Moves [17]) *Let $L \in \mathcal{B}(A^*)$ be an event log over the activities A, and let M be a model where T is the set of tasks of the model, $A_v(M)$ is the set of observable activities of M, and l is the labeling function between tasks and observable activities in M. For the sake of clarity, we abuse the notation writing $l(t) = \tau$ if $t \notin dom(l)$, i.e., if t is an invisible task.*

- *$(a_L, (a_M, t))$ is a move, where $a_L \in A \cup \gg$, $(a_M, t) \in (A_v(M) \cup \tau \times T) \cup \gg$, and $l(t) = a_M$.*
- *$(a, (a, t))$ is a synchronous move (also called move in both), where $a \in A$, $t \in T$, and $l(t) = a$.*
- *$(a, \gg)$ is a move on log, where $a \in A$.*
- *$(\gg, (a, t))$ is a move on model, where $(a, t) \in (A_v(M) \cup \tau \times T)$, and $l(t) = a$.*
- *A legal move is a move on log, a move on model, or a synchronous move. Otherwise, it is an illegal move. A_{LM} denotes the set of possible legal moves between the model M and log L.*

Given a sequence of moves γ, $row_L^{\gg}(\gamma)$ denotes the sequence of log activities in γ, i.e., the first element. Similar, $row_M^{\gg}(\gamma)$ and $row_T^{\gg}(\gamma)$ denote the sequence of model activities and tasks, respectively. row_L, row_M and row_T denote the projection of sequences of activities in the log, model and tasks, filtering $\gg$.

Definition 9.2 (Alignment [17]) *Let $\sigma_L \in L$ be a log trace and $\sigma_M \in \phi_t(M)$ a complete task sequence of model M. An alignment of σ_L and σ_M is a sequence of moves $\gamma \in A_{LM}^*$ such that the sequence of log activities (ignoring $\gg$) yields σ_L, and the sequence of model tasks (ignoring $\gg$) yields σ_M, i.e., $row_L(\gamma) = \sigma_L$ and $row_T(\gamma) = \sigma_M$.*

Let us consider a medical process for an oncological treatment in a hospital – this process will be used as running example during this chapter. Model M in Figure 9.2 represents a possible model for this medical process. Assuming that the end state of the model is reached when place *end* in the model contains exactly one token, the model represents an infinite set of complete activity sequences, e.g., $\langle a,b,c,d \rangle$, $\langle a,c,b,d \rangle$, $\langle a,b,c,e \rangle$, $\langle a,c,b,e \rangle$, $\langle a,f,g,h \rangle$, $\langle a,b,i,c,b,e \rangle$. Given an unfitting trace $\sigma_L = \langle a,b,d,e \rangle$, Figure 9.3 shows some possible alignments between σ_L and M.

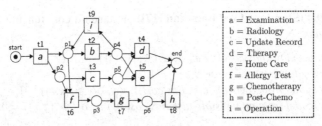

Fig. 9.2: Model for a medical process, used as running example on this chapter.

$$\gamma_1 = \begin{array}{|c|c|c|c|c|}\hline a & \gg & b & d & e \\\hline a & c & b & \gg & e \\\hline t_1 & t_3 & t_2 & & t_5 \\\hline\end{array} \quad \gamma_2 = \begin{array}{|c|c|c|c|c|}\hline a & b & \gg & d & e \\\hline a & b & c & \gg & e \\\hline t_1 & t_2 & t_3 & & t_5 \\\hline\end{array} \quad \gamma_3 = \begin{array}{|c|c|c|c|c|}\hline a & \gg & b & d & e \\\hline a & c & b & d & \gg \\\hline t_1 & t_3 & t_2 & t_4 & \\\hline\end{array}$$

$$\gamma_4 = \begin{array}{|c|c|c|c|c|}\hline a & b & \gg & d & e \\\hline a & b & c & d & \gg \\\hline t_1 & t_2 & t_3 & t_4 & \\\hline\end{array} \quad \gamma_5 = \begin{array}{|c|c|c|c|c|}\hline a & b & d & \gg & e \\\hline a & b & \gg & c & e \\\hline t_1 & t_2 & & t_3 & t_5 \\\hline\end{array} \quad \gamma_6 = \begin{array}{|c|c|c|c|c|c|c|}\hline a & \gg & \gg & \gg & b & d & e \\\hline a & f & g & h & \gg & \gg & \gg \\\hline t_1 & t_6 & t_7 & t_8 & & & \\\hline\end{array}$$

Fig. 9.3: Some alignments between trace $\sigma_L = \langle a,b,d,e \rangle$ and the model M in Figure 9.2.

The moves are represented vertically, e.g., the second move of γ_1 is $(\gg,(c,t_3))$, indicating that the model performs t_3 while the log does not make any move. Note that after removing $\gg$, the projections of all moves in the model are by definition complete task sequences allowed by the model. This property is not always ensured by other conformance checking approaches. For example, given a trace and a process model, when using the approach in [77], the so-called *missing tokens* are added to allow activities that occur in the trace but are not supposed to occur according to the model. The addition of such missing tokens introduces extra behavior that is not allowed in the original process model, thus overestimating its behavior.

In order to compare alignments and select the most appropriate one, *costs* are associated to undesirable moves and the alignment with the lowest total costs is selected. To quantify the costs of an alignment, a *cost function* δ is defined.

Definition 9.3 (Cost of alignment [17]) *The* cost function $\delta : A_{LM} \rightarrow \mathbb{N}$ *assigns costs to legal moves. The* cost *of an alignment* $\gamma \in A_{LM}^*$ *is the sum of all costs, i.e.,*
$$\delta(\gamma) = \sum_{(x,y)\in\gamma} \delta(x,y).$$

Different scenarios may require different cost functions. The costs may depend on the nature of the activity, e.g., skipping a payment may be more severe than sending an email. Moreover, the severity assumed for a move on log and a move on model may be different, e.g., a system with constant recording problems should be more tolerant with activities skipped on the log. Abstracting from particular cases, we can define a *standard cost function* that assigns unit costs to moves in log or moves on model only.

Definition 9.4 (Standard Cost Function [17]) *A standard cost function δ_S is defined such that:*

- *Synchronous move has cost 0, i.e., $\delta_S(x,(x,t)) = 0$ for all $x \in A$.*
- *Move on log has cost 1, i.e., $\delta_S(x,\gg) = 1$.*
- *Move on model from a visible task has cost 1, i.e., $\delta_S(\gg,(x,t)) = 1$.*
- *Move on model from an invisible task has cost 0, i.e., $\delta_S(\gg,(\tau,t)) = 0$.*

Using the standard cost function, the cost of alignment γ_1 is $\delta_S(\gamma_1) = \delta_S(a,(a,t_1))$ $+ \delta_S(\gg,(c,t_3)) + \delta_S(b,(b,t_2)) + \delta_S(d,\gg) + \delta_S(e,(e,t_5)) = 0+1+0+1+0 = 2$. Note that the function returns the number of mismatches in the alignment. On the other hand, $\delta_S(\gamma_6) = 6$. Hence, we conclude that γ_1 is close to the log trace $\sigma_L = \langle a,b,d,e \rangle$ than γ_6.

Given a trace from an event log and a process model, we are interested in an activity sequence from the model that is most similar to the trace, i.e., the *optimal alignment*.

Definition 9.5 (Optimal Alignments [17]) *We define the set of alignments $\Gamma_{\sigma_L,M} = \{\gamma \in A_{LM}{}^* \mid \gamma$ is an alignment between σ_L and $M\}$ to be all possible alignments between σ_L and M. Accordingly, we define the set of optimal alignments as the set of all alignments with minimum cost, i.e., $\Gamma_{\sigma_L,M}^o = \{\gamma \in \Gamma_{\sigma_L,M} \mid \forall_{\gamma' \in \Gamma_{\sigma_L,M}} \delta(\gamma) \le \delta(\gamma')\}$.*

It is easy to see that there can be more than one optimal alignment between a trace and a model. For example, $\{\gamma_1,\gamma_2,\gamma_3,\gamma_4,\gamma_5\}$ is the set of optimal alignments between the trace $\sigma_L = \langle a,b,d,e \rangle$ and the model M.

By definition, the task component of all alignments yields a complete task sequence of the model. Thus, given an optimal alignment γ between σ_L and M, $row_T(\gamma)$ provides a complete tasks sequence that both perfectly fits M and is closest to σ_L. In the running example, $row_T(\gamma_1) = \langle t_1,t_3,t_2,t_5 \rangle$ is one of the complete task sequences of M that is most similar to trace $\langle a,b,d,e \rangle$.

Given a log and a model, constructing all optimal alignments between all traces in the log and the model is computationally expensive [18, 19]. Thus, there are cases where computing all optimal alignments between traces and process models may not always be feasible in practice. Hence, instead of computing all optimal alignments between traces in the log and the model to obtain insights into deviations, one may also compute just some representative optimal alignments for each trace. In this chapter, we consider three approaches: *one* optimal alignment per trace, *all* optimal approaches, and a set of *representative* optimal alignments. We define three functions that provide optimal alignments between traces in the log and the model:

- $\Lambda_M^* : A_L^* \to \mathscr{P}(A_{LM}{}^*)$ returns *all optimal alignments* between traces of L and M, such that for all $\sigma_L \in L, \Lambda_M^*(\sigma_L) = \Gamma_{\sigma_L,M}^o$,
- $\Lambda_M^1 : A_L^* \to A_{LM}{}^*$ returns *one optimal alignment* between traces of L and M, such that for all $\sigma_L \in L, \Lambda_M^1(\sigma_L) \in \Gamma_{\sigma_L,M}^o$, and
- $\Lambda_M^R : A_L^* \to \mathscr{P}(A_{LM}{}^*)$ returns *representatives of optimal alignments* between traces of L and M, such that for all $\sigma_L \in L, \Lambda_M^R(\sigma_L) \subseteq \Gamma_{\sigma_L,M}^o$.

$$\gamma_7 = \begin{array}{|c|c|c|c|} \hline a & \gg & \gg & \gg \\ \hline a & f & g & h \\ \hline t_1 & t_6 & t_7 & t_8 \\ \hline \end{array} \quad \gamma_8 = \begin{array}{|c|c|c|c|} \hline a & \gg & \gg & \gg \\ \hline a & b & c & d \\ \hline t_1 & t_2 & t_3 & t_4 \\ \hline \end{array} \quad \gamma_9 = \begin{array}{|c|c|c|c|} \hline a & \gg & \gg & \gg \\ \hline a & c & b & d \\ \hline t_1 & t_3 & t_2 & t_4 \\ \hline \end{array}$$

$$\gamma_{10} = \begin{array}{|c|c|c|c|} \hline a & \gg & \gg & \gg \\ \hline a & c & b & e \\ \hline t_1 & t_3 & t_2 & t_5 \\ \hline \end{array} \quad \gamma_{11} = \begin{array}{|c|c|c|c|} \hline a & \gg & \gg & \gg \\ \hline a & b & c & e \\ \hline t_1 & t_2 & t_3 & t_5 \\ \hline \end{array}$$

Fig. 9.4: All optimal alignments between trace $\sigma_L = \langle a \rangle$ and the model M in Figure 9.2.

In [22, 18, 19] various approaches to obtain an optimal alignment between a trace and a model with respect to different cost functions are investigated. Given a trace σ_L of L and a model M, if there are multiple optimal alignments, Λ_M^1 chooses one of them according to other external criteria. With our previous example, suppose that Λ_M^1 selects an alignment that has the longest consecutive occurrence of synchronous moves in the beginning, $\Lambda_M^1(\sigma_L) = \gamma_4$.

In [18, 19], an $A^\star$-based algorithm is proposed to compute one optimal alignment between a trace and a model. The same algorithm can be extended to provide more than one optimal alignment between them. Given a trace σ_L of L and a model M, the algorithm constructs one optimal alignment by computing a shortest path from the initial to the final state of the state space of the synchronous product between σ_L and M. It is shown in [19] that all shortest paths from the initial to the final state of the state space yield an optimal alignment. For each state in the state space, the algorithm records a shortest path from the initial state to reach this state and thus, becomes the *representative* of all other shortest paths from the initial state to the state. An optimal alignment is constructed from a shortest path from the initial state to the final state that is also representing all other shortest paths that connect the same pair of states. By recording all represented shortest paths during state space exploration for each state, we can obtain all shortest paths from the initial to the final state of the state space (i.e., obtain *all optimal alignments*). Different representatives may represent different number of optimal alignments. Given a representative $\gamma \in \Lambda_M^R(\sigma_L), rep_M(\gamma)$ denotes the number of optimal alignments represented by γ. Furthermore, due to possible pruning of state space, the total number of represented optimal alignments by the representatives is a lower bound of the total number of all optimal alignments, i.e., $\sum_{\gamma \in \Lambda_M^R(\sigma_L)} rep_M(\gamma) \leq |\Gamma_{\sigma_L,M}^o|$. The interested reader is referred to [18, 19, 17] for details on the constructed state space with the $A^\star$-based algorithm approach.

Take for example a trace $\sigma_L = \langle a \rangle$. All optimal alignments between the trace and the medical model M are shown in Figure 9.4. Given a possible function Λ^R, $\Lambda^R(\sigma_L) = \{\gamma_7, \gamma_9, \gamma_{10}\}$ where $rep_M(\gamma_7) = 1$ (γ_7 represents $\{\gamma_7\}$), $rep(\gamma_9) = 2$ (γ_9 represents $\{\gamma_8, \gamma_9\}$), and $rep(\gamma_{10}) = 2$ (γ_{10} represents $\{\gamma_{10}, \gamma_{11}\}$).

For simplicity, in the remainder we omit the model notation M in functions $\Lambda_M^\star$, Λ_M^1, Λ_M^R, and rep_M whenever the context is clear. Note that in cases where a process model has duplicate tasks (more than one task to represent an activity) or invisible tasks (tasks whose execution are not logged), approaches to construct alignments (e.g., [22, 18]) keep the mapping from all model moves to the tasks they correspond to. Hence, given an alignment of a trace and such models, we know exactly which task is executed for each model move. We refer to [22, 18] for further details on how such mapping is constructed.

9.3 Precision based on Alignments

The technique described in the previous section provides optimal alignments for each trace in the log. This section presents a technique to compute *precision* based on the use of these optimal alignments. Like the approach on Chapter 5, the behavior observed in the log is used to traverse the modeled behavior, detecting escaping arcs for possible points of imprecision. However, whereas in Chapter 5 is based on model replay directly from the log, the approach presented here uses the alignments as a more faithful representation of the observed behavior. The advantages are manifold. First of all, traces in the log do not need to be completely fitting. In Chapter 5, the non-fitting parts are simply ignored. For most real-life situations, this implies that only a fraction of the event log can be used for computing precision. Second, the existence of indeterminism in the model poses no problems when using the alignments. Finally, the use of alignments instead of log-based model replay improves the robustness of conformance checking. The remainder of this section is devoted to explain how precision can be calculated from the alignments. In particular, we consider the precision checked from *one* alignment, *all* alignments, and *representative* alignments. To illustrate the three approaches, in the remainder of the section we use the following running example: the model M shown in Figure 9.2 and the log $L = [\sigma_1, \sigma_2, \sigma_3, \sigma_4, \sigma_5]$, containing the 5 traces that appear in in Table 9.1. The table also provides the optimal alignments for the traces in L.

Freq	Trace	Optimal Alignment

1   $\sigma_1 = \langle a \rangle$

$$\gamma_{1a} = \begin{array}{|c|c|c|c|} \hline a & \gg & \gg & \gg \\ \hline a & f & g & h \\ \hline t_1 & t_6 & t_7 & t_8 \\ \hline \end{array}$$

$$\gamma_{1b} = \begin{array}{|c|c|c|c|} \hline a & \gg & \gg & \gg \\ \hline a & b & c & d \\ \hline t_1 & t_2 & t_3 & t_4 \\ \hline \end{array} \qquad \gamma_{1c} = \begin{array}{|c|c|c|c|} \hline a & \gg & \gg & \gg \\ \hline a & c & b & d \\ \hline t_1 & t_3 & t_2 & t_4 \\ \hline \end{array}$$

$$\gamma_{1d} = \begin{array}{|c|c|c|c|} \hline a & \gg & \gg & \gg \\ \hline a & c & b & e \\ \hline t_1 & t_3 & t_2 & t_5 \\ \hline \end{array} \qquad \gamma_{1e} = \begin{array}{|c|c|c|c|} \hline a & \gg & \gg & \gg \\ \hline a & b & c & e \\ \hline t_1 & t_2 & t_3 & t_5 \\ \hline \end{array}$$

1   $\sigma_2 = \langle a,b,c,d \rangle$

$$\gamma_2 = \begin{array}{|c|c|c|c|} \hline a & b & c & d \\ \hline a & b & c & d \\ \hline t_1 & t_2 & t_3 & t_4 \\ \hline \end{array}$$

1   $\sigma_3 = \langle a,c,b,e \rangle$

$$\gamma_3 = \begin{array}{|c|c|c|c|} \hline a & c & b & e \\ \hline a & c & b & e \\ \hline t_1 & t_3 & t_2 & t_5 \\ \hline \end{array}$$

1   $\sigma_4 = \langle a,f,g,h \rangle$

$$\gamma_4 = \begin{array}{|c|c|c|c|} \hline a & f & g & h \\ \hline a & f & g & h \\ \hline t_1 & t_6 & t_7 & t_8 \\ \hline \end{array}$$

1   $\sigma_5 = \langle a,b,i,b,c,d \rangle$

$$\gamma_5 = \begin{array}{|c|c|c|c|c|c|} \hline a & b & i & b & c & d \\ \hline a & b & i & b & c & d \\ \hline t_1 & t_2 & t_9 & t_2 & t_3 & t_4 \\ \hline \end{array}$$

Table 9.1: Optimal alignments of log $[\sigma_1, \sigma_2, \sigma_3, \sigma_4, \sigma_5]$ and the medical model M of Figure 9.2

9.4 Precision from 1-Alignment

Like Chapter 5, precision is estimated by confronting model and log behavior: escaping arcs between the model and the log (i.e., situations where the model allows more behavior than the one reflected in the log) are detected by juxtaposing behavior observed in the log and the one allowed by the model. This juxtaposition is done in terms of an automaton: first, an automaton is built from the alignments. Then, the automaton is enhanced with behavioral information of the model. Finally, the enhanced automaton is used to compute the precision. In order to build the automaton, observed behavior must be determined in terms of model perspective, i.e., we consider the optimal alignments of each trace in the log for this purpose. For example, given the running example L and M, the trace σ_1 has 5 optimal alignments, $\gamma_{1a}, \gamma_{1b}, \gamma_{1c}, \gamma_{1d}, \gamma_{1e}\}$, shown in Table 9.1. However, in 1-alignment only one alignment is considered. For this example, we assume that the alignment assigned to σ_1 by Λ^1 based on an external criterion corresponds to γ_{1a}, i.e., $\Lambda^1(\sigma_1) = \gamma_{1a}$. On the other hand, traces $\sigma_2 \ldots \sigma_5$ are perfectly fitting and have only one optimal alignment containing only synchronous moves. Given an alignment γ, in order to build the automaton, we only consider the projection of tasks moves, i.e., $row_T(\gamma)$. In this example, the sequences used as observed behavior are $\langle t_1, t_6, t_7, t_8 \rangle$, $\langle t_1, t_2, t_3, t_4 \rangle$, $\langle t_1, t_3, t_2, t_5 \rangle$, $\langle t_1, t_6, t_7, t_8 \rangle$ and $\langle t_1, t_2, t_9, t_2, t_3, t_4 \rangle$. We use $row_T(\Lambda^1)_L$ to denote the application of function row_T on all the alignments provided by the functions Λ^1 for the traces in log L. We can omit the subindex L whenever the context is clear. Note that, by definition, any alignment projection $row_T(\gamma)$ is a valid complete firing sequence of the model.

Similar to Chapter 5, the automaton is built considering all the prefixes for the sequences in $row_T(\Lambda^1)$ as the states. For instance, given a sequence $\langle t_1, t_2, t_3, t_4 \rangle$ resulting of $row_T(\Lambda^1)(\sigma_2)$, the states considered are $\langle \rangle$, $\langle t_1 \rangle$, $\langle t_1, t_2 \rangle$, $\langle t_1, t_2, t_3 \rangle$ and $\langle t_1, t_2, t_3, t_4 \rangle$. We denote as $\bullet(row_T(\gamma))$ the set of *prefixes of the tasks sequence* of the alignment γ and as $\bullet(row_T(\Lambda^1))$ the multiset of *prefixes of the the tasks sequences* of all alignments in Λ^1.

Definition 9.6 (Prefix Automaton of the 1-Alignment) *Let $L \in \mathscr{B}(A^*)$ be an event log, let M be a model with tasks T, and let $row_T(\Lambda^1)$ be the alignments between them projected on the model tasks. We define the* prefix *automaton of the 1-Alignment $\mathscr{A}_{\Lambda^1 M} = (S, T, \nearrow, \omega, s_0)$ such that:*

- *the set of* states *corresponds to the set of prefixes of the alignments projected on the model tasks, i.e., $S = \{\sigma | \sigma \in \bullet(row_T(\Lambda^1))\}$.*
- *the set of* labels *corresponds to the set of tasks T.*
- *the arcs $\nearrow \subseteq (S \times T \times S)$ define the concatenation between states and tasks, i.e., $\nearrow = \{(\sigma, t, \sigma \cdot \langle t \rangle) | \sigma \in S \wedge \sigma \cdot \langle t \rangle \in S\}$.*
- *the function that determines the* weight *of a state is determined by the number of occurrences of the state in the multiset of prefixes of the tasks sequences, i.e., $\omega(\sigma) = \bullet(row_T(\Lambda^1))(\sigma)$ for all $\sigma \in S$.*
- *the initial state s_0 corresponds with the empty prefix $\langle \rangle$.*

Figure 9.5 shows the resulting automata for the running example L using the function Λ^1 (only the white states). For example, the weight of the state $\langle t_1 \rangle$ is greater than the weight of $\langle t_1, t_3 \rangle$ because there are more tasks sequences with the prefix $\langle t_1 \rangle$ (all 5 sequences), than the ones with prefix $\langle t_1, t_3 \rangle$ (only the sequence $\langle t_1, t_3, t_2, t_5 \rangle$ contains that prefix).

Once the observed behavior has been determined in terms of an automaton, the confrontation with the actual modeled behavior is required in order to determine the precision. For each state of the automaton, we compute its set of *modeled tasks*, i.e., possible direct successor tasks according to the model (*mod*), and then compare it with the set of *observed tasks*, i.e., tasks really executed in the log (*obs*)(cf. Definition 5.4). Let us consider, for example, state $\langle t_1, t_2, t_3 \rangle$ of automaton in Figure 9.5.

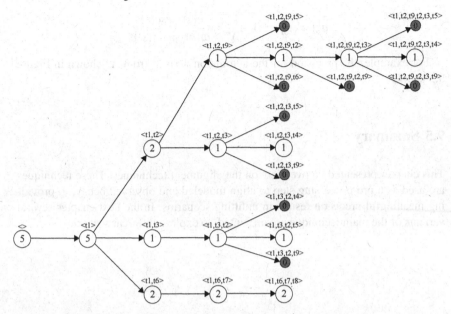

Fig. 9.5: Automaton from 1-alignments between model M and log L.

The set of observed tasks of the state is $obs(\langle t_1, t_2, t_3 \rangle) = \{t_4\}$, i.e., for all traces with prefix $\langle t_1, t_2, t_3 \rangle$, their direct successor is only t_4. The set of modeled tasks for the state is $mod(\langle t_1, t_2, t_3 \rangle) = \{t_4, t_5, t_9\}$ because after performing the sequence of tasks $\langle t_1, t_2, t_3 \rangle$, the model allows to do t_4, t_5 or t_9. Note that, by definition of alignment, $obs(s) \subseteq mod(s)$, i.e., the set of executed tasks of a given state is always a subset of all available tasks according to the model.

The arcs that are modeled according to the model, but do not occur in the event log according to the alignments, are used to collect the *escaping arcs* of the system, i.e., arcs that escapes from the observed behavior. The tasks on the escaping arcs and the states reached are called escaping tasks and escaping states respectively. In Figure 9.5 the escaping states are in color. For example, the escaping tasks of the state $\langle t_1, t_2, t_3 \rangle$ are $\{t_4, t_5, t_9\} \setminus \{t_4\} = \{t_5, t_9\}$. The computation and analysis of these escaping arcs are the cornerstone of the precision checking technique presented in this book. All identified escaping arcs can be analyzed and further used to correct the model and make it more precise. Furthermore, in order to globally estimate precision, these escaping arcs in turn are weighted and normalized defining a metric to measure precision called *1-align precision* metric.

Definition 9.7 (1-Align Precision metric) *Let $\mathscr{A}_{\Lambda^1 M} = (S, T, \nearrow, \omega, s_0)$ be the prefix automaton of the alignments in Λ^1 enhanced with the behavior of the model M. The metric* 1-Align Precision *estimates the precision of the system comparing, for each state in S, the number of escaping arcs with the number of allowed arcs. The numbers are weighted according to the importance of the state. Formally:*

$$a_p^1(\mathscr{A}_{\Lambda^1 M}) = 1 - \frac{\sum_{s \in S} \omega(s) \cdot |esc(s)|}{\sum_{s \in S} \omega(s) \cdot |mod(s)|}$$

For example, the precision for the automaton derived from Λ^1 shown in Figure 9.5 is 0.79.

9.5 Summary

This chapter presented an overview on the alignment techniques. These techniques are used as a pre-processing step to align modeled and observed behavior, providing meaningful precision results in unfitting scenarios. In the next chapter several variants of the main technique are presented, to explore different scenarios.

Chapter 10
Alternative and Variants to Handle non-Fitness

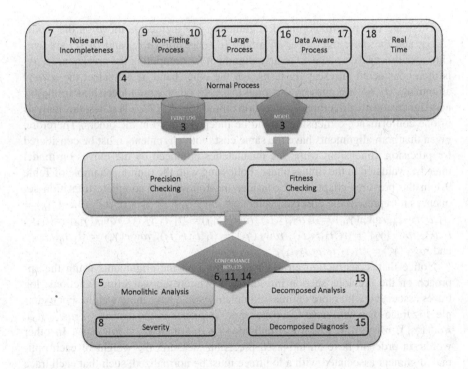

The previous chapter presented the use of the alignment technique to compute precision. This chapter provides alternatives to the basic alignment approach to analyze different types of scenarios. In particular, it concentrates on the use of several optimal alignments and the directionality of those alignments. In the next chapter, the approach to check precision based on alignments is put into practice.

© Springer International Publishing AG 2016
J. Munoz-Gama: Conf. Check. ... in Process Mining, LNBIP 270, pp. 97–106, 2016.
DOI: 10.1007/978-3-319-49451-7_10

10.1 Precision from All-Alignment

As experiments show, the use of one alignment is an effective and efficient alternative for precision checking. However, the approach relies on selecting one of optimal alignments, and therefore, it may detect more escaping arcs. Let us consider the model in Figure 10.1, and a log with only one trace $\langle a, d \rangle$.

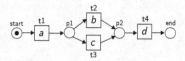

Fig. 10.1: Model with two optimal alignments for the trace $\langle a, d \rangle$, one with move on the model (b, t_2), and the other with one move on model (c, t_3).

There are two possible optimal alignments: one includes a move on the model (b, t_2), and the other one move on model (c, t_3), while (a, t_1) and (d, t_4) are synchronous moves. If we select the first alignment for computing the precision, $\langle t_1, t_3 \rangle$ is then considered an escaping tasks. On the other hand, if we select the second alignment, $\langle t_1, t_2 \rangle$ becomes an escaping tasks. In other words, decisions in the fitness dimension are affecting the precision dimension. As [77, 73] states, the analysis of one conformance dimension should be independent from the others. Therefore, given that both alignments have the same cost, both alignments must be considered for precision dimension, while the mismatches produced by the moves on model must be evaluated in the fitness phase. Following with the running example of Table 9.1 in the previous chapter, we consider the following 9 complete transition sequences to represent the observed behavior: $row_T(\gamma_{1a}) = \langle t_1, t_6, t_7, t_8 \rangle$, $row_T(\gamma_{1b}) = \langle t_1, t_2, t_3, t_4 \rangle$, $row_T(\gamma_{1c}) = \langle t_1, t_3, t_2, t_4 \rangle$, $row_T(\gamma_{1d}) = \langle t_1, t_3, t_2, t_5 \rangle$, $row_T(\gamma_{1e}) = \langle t_1, t_2, t_3, t_5 \rangle$, $row_T(\gamma_2) = \langle t_1, t_2, t_3, t_4 \rangle$, $row_T(\gamma_3) = \langle t_1, t_3, t_2, t_5 \rangle$, $row_T(\gamma_4) = \langle t_1, t_6, t_7, t_8 \rangle$, and $row_T(\gamma_5) = \langle t_1, t_2, t_9, t_2, t_3, t_4 \rangle$,

Notice that, constructing an automaton from all the alignments using the approach on the previous section introduces a bias on the weighting functions: log traces associated with more optimal alignments would have more weight. For example, log trace σ_1 would have five times more influence in the measure – $row_T(\gamma_{1a})$, $row_T(\gamma_{1b})$, $row_T(\gamma_{1c})$, $row_T(\gamma_{1d})$ and $row_T(\gamma_{1e})$ – than σ_2 – $row_T(\gamma_2)$. In other words, in order to have an unbiased precision measure, the weight of each optimal alignment associated with a log trace must be normalized, such that each trace of the log has the same importance within the observed behavior. For that, we define the *weight* of an alignment, and we redefine the weight of the states in the prefix automaton.

Definition 10.1 (Weight of Alignment) *Let M be a model, let L be an event log, let σ_L be a trace in L, let $L(\sigma_L)$ be the frequency of σ_L, and let $\gamma \in \Lambda_M^*(\sigma_L)$ be one of the optimal alignments between σ_L and M. The weight of γ is defined as $\omega(\gamma) = L(\sigma_L) \cdot 1/|\Lambda_M^*(\sigma_L)|$, i.e., the weight is split equally among all the alignments* *of the log trace, taking into account the frequency of the trace withing the log. Given the prefix automaton $\mathscr{A}_{\Lambda^*M} = (S, T, \nearrow, \omega, s_0)$, the weight of a state $s \in S$ is defined as:*

$$\omega(s) = \sum_{\forall \gamma \in \Lambda^*} \omega(\gamma) \ \textit{if s is a prefix of } row_T(\gamma) \ (\textit{or 0 otherwise})$$

For example, the weight of the alignment γ_{1a} of trace σ_1 is $1 \cdot 1/5 = 0.2$, while the weight of γ_2 is $1 \cdot 1/1 = 1$. Figure 10.2 shows the resulting automata for the running example L and M using the function Λ^*. For example, the weight of the state $\langle t_1, t_6 \rangle$ is 1 from γ_4 plus 0.2 from γ_{1a}, i.e., 1.2.

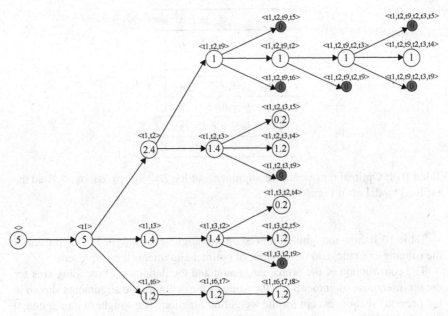

Fig. 10.2: Automaton from all-alignments between model M and log L.

As it is done with 1-align precision metric, the all-align metric is formalized as:

Definition 10.2 (All-Align Precision metric) *Let $\mathscr{A}_{\Lambda^*M} = (S, T, \nearrow, \omega, s_0)$ be the prefix automaton of the alignments in Λ^* enhanced with the behavior of the model M. The metric* All-Align Precision *estimates the precision of the system comparing, for each state in S, the number of escaping arcs with the number of allowed arcs. The numbers are weighted according to the importance of the state. Formally:*

$$a_p^*(\mathscr{A}_{\Lambda^*M}) = 1 - \frac{\sum_{s \in S} \omega(s) \cdot |esc(s)|}{\sum_{s \in S} \omega(s) \cdot |mod(s)|}$$

10.2 Precision from Representative-Alignment

Given a log trace and a process model, Λ^* provides all optimal alignments. However, as shown in [20], it is an expensive option in terms of computation time. The use of only one alignment per trace (i.e., Λ^1) solves this issue in cases where time is a priority, but it may sacrifice accuracy. As a trade-off between time and accuracy, in this section we propose precision measurement based on *representatives* of all optimal alignments Λ^R (cf. Section 9.2). In this section, we revisit the precision measurement to include this notion.

Freq	Trace	Λ^R	rep
1	$\sigma_1 = \langle a \rangle$	γ_{1a}	1
		γ_{1c}	2
		γ_{1d}	2
1	$\sigma_2 = \langle a,b,c,d \rangle$	γ_2	1
1	$\sigma_3 = \langle a,c,b,e \rangle$	γ_3	1
1	$\sigma_4 = \langle a,f,g,h \rangle$	γ_4	1
1	$\sigma_5 = \langle a,b,i,b,c,d \rangle$	γ_5	1

Table 10.1: Optimal representative alignments of log $L = [\sigma_1, \sigma_2, \sigma_3, \sigma_4, \sigma_5]$ and the medical model M of Figure 9.2

Table 10.1 show the optimal representative alignments assigned to each trace of the running example, and the number of optimal alignments they represent.

The construction of the prefix automaton and the definition of escaping arcs for the representative alignments are the same as the or for all the alignments shown in the previous section, except for the weighting function. The weight of an alignment γ needs to be proportional to the number of alignments represented by γ, i.e., $rep(\gamma)$.

Definition 10.3 (Weight of Alignment) *Let M be a model, let L be an event log, let σ_L be a trace in L, let $L(\sigma_L)$ be the frequency of σ_L, and let $\gamma \in \Lambda_M^R(\sigma_L)$ be one of the optimal representative alignments between σ_L and M, where $rep(\gamma)$ are the alignments represented by γ. The weight of γ is defined as:*

$$\omega(\gamma) = L(\sigma_L) \cdot rep(\gamma) / \sum_{\gamma' \in \Lambda^R(\sigma_L)} rep(\gamma')$$

For instance, let us consider the optimal representative alignment γ_{1c} for the log trace σ_1. The number of alignments represented by γ_{1c} is $rep(\gamma_{1c}) = 2$. The total number of optimal alignments represented by the representative alignments associated with σ_1 is $\sum_{\gamma' \in \Lambda^R(\sigma_1)} rep(\gamma') = 5$. Hence, the weight $\omega(\gamma_{1c}) = 1 \cdot 2/5 = 0.4$. On the other hand, let us consider γ_5, the only optimal alignment associated with σ_5. The representative alignment γ_5 represents 1 optimal alignment. Since the number of all optimal alignments represented is $\sum_{\gamma' \in \Lambda^R(\sigma_5)} rep(\gamma') = 1$, the weight of γ_5 is $\omega(\gamma_5) = 1 \cdot 1/1 = 1$.

Figure 10.3 shows the resulting automata for the running example L and M using the function Λ^R. For example, the weight of the state $\langle t_1, t_3 \rangle$ is 1 from γ_3, plus 0.4 from γ_{1c} (represents 2 alignments of 5), plus 0.4 from γ_{1d} (represents 2 alignments of 5), i.e., 1.8.

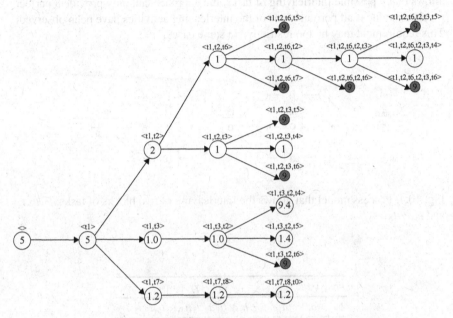

Fig. 10.3: Automaton from representative-alignments between model M and log L.

As it is done with 1-align and all-align precision metric, the rep-align metric is formalized as:

Definition 10.4 (Rep-Align Precision metric) *Let $\mathscr{A}_{\Lambda^R M} = (S, T, \nearrow, \omega, s_0)$ be the prefix automaton of the alignments in Λ^R enhanced with the behavior of the model M. The metric All-Align Precision estimates the precision of the system comparing, for each state in S, the number of escaping arcs with the number of allowed arcs. The numbers are weighted according to the importance of the state. Formally:*

$$a_p^R(\mathscr{A}_{\Lambda^R M}) = 1 - \frac{\sum_{s \in S} \omega(s) \cdot |esc(s)|}{\sum_{s \in S} \omega(s) \cdot |mod(s)|}$$

Note that there can be more than one ways to compute representative alignments from a given model and a trace. Given an event log and a model, the selection of representative alignments between each trace in the log and the model obviously influences the automata that can be constructed between the log and the model.

10.3 Abstractions for the Precision based on Alignments

The approach presented in Section 9.3 uses the prefixes of complete tasks sequences to represent states of the automaton. This implies that given a complete tasks sequence σ, other sequences with slightly different permutation of tasks are placed in different branches of constructed automaton than σ. Given a process model that allows many possible interleaving of tasks, the approach can only provide a perfect precision value if all permutations of the interleaving activities have been observed. This requirement may be too restrictive in some cases.

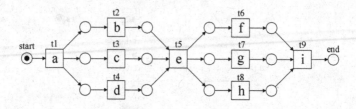

Fig. 10.4: Process model that allows the interleaving of two blocks of tasks: t_2, t_3, t_4 and t_6, t_7, t_8.

Freq	Trace	$row_T(\gamma)$
1	$\langle a,b,c,d,e,f,g,h,i \rangle$	$\langle t_1,t_2,t_3,t_4,t_5,t_6,t_7,t_8,t_9 \rangle$
1	$\langle a,b,d,c,e,f,h,g,i \rangle$	$\langle t_1,t_2,t_4,t_3,t_5,t_6,t_8,t_7,t_9 \rangle$
1	$\langle a,c,b,d,e,g,f,h,i \rangle$	$\langle t_1,t_3,t_2,t_4,t_5,t_7,t_6,t_8,t_9 \rangle$
1	$\langle a,c,d,b,e,g,h,f,i \rangle$	$\langle t_1,t_3,t_4,t_2,t_5,t_7,t_8,t_6,t_9 \rangle$
1	$\langle a,d,b,c,e,h,f,g,i \rangle$	$\langle t_1,t_4,t_2,t_3,t_5,t_8,t_6,t_7,t_9 \rangle$
1	$\langle a,d,c,b,e,h,g,f,i \rangle$	$\langle t_1,t_4,t_3,t_2,t_5,t_8,t_7,t_6,t_9 \rangle$

Table 10.2: Event log for the model in Figure 10.4

For example, let us consider the process model in Figure 10.4 and the log in Table 10.2. The model allows for the interleaved execution of t_2, t_3 and t_4. This behavior is also observed in the log, containing all possible permutations of t_2, t_3 and t_4. The model also allows the interleaving of t_6, t_7 and t_8, and all possible permutations of t_6, t_7 and t_8 are also observed in the log. One may expect a perfect precision of 1 for such model and log. However, given the presented approach, the precision is 0.8. The automaton of Figure 10.5 shows the escaping arcs detected. Notice that prefix $\langle t_1,t_2,t_3 \rangle$ of $\langle t_1,t_2,t_3,t_4,t_5,t_6,t_7,t_8,t_9 \rangle$ and prefix $\langle t_1,t_3,t_2 \rangle$ of $\langle t_1,t_3,t_2,t_4,t_5,t_7,t_6,t_8,t_9 \rangle$ represent two different states even when the executed tasks and their frequency in both prefixes are the same. For the given example, the minimum number of traces necessary to reach a precision of 1 is 36. This number increases exponentially with the increasing degree of concurrency of the considered model. In such cases, some level of abstraction in the way states are represented is desirable. In Section 10.3.1 we propose an approach to abstract from the order of the tasks to compute the precision, dealing with the possible incompleteness of the log.

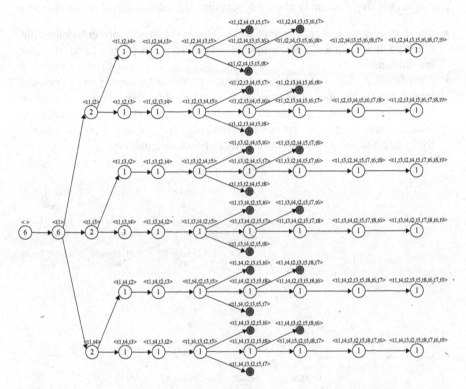

Fig. 10.5: Precision automaton and escaping arcs between the model in Figure 10.4 and the log in Table 10.2.

Moreover, notice that the automaton is constructed considering the prefixes of the complete tasks sequences. However, this may introduce a bias on the *direction*, i.e., the tasks executed in the *past* are the ones that determine the current state. A reasonable alternative is to consider the *future* tasks to determine the current state. In Section 10.3.2 we propose to use the future to construct the automaton, in order to deal with the possible bias produced by the direction used to compute precision.

10.3.1 Abstraction on the Order

In [14], the authors describe an approach to extract a transition system from the log. The proposed approach considers a set of possible abstractions and filters on the trace in order to determine the states of the transition system. In particular, they propose the use of sets, multisets, and sequences of activities as state representations. In a similar way, in this section we propose two possible state representations for precision checking that can be chosen depending on the desired level of abstractions:

- *Ordered*: A state is a *sequence* of tasks. This is the same representation as the one used in Section 9.3. For example, the states for prefix $\langle t_1, t_2, t_3 \rangle$ and $\langle t_1, t_3, t_2 \rangle$ are different.
- *Unordered*: A state is a *multiset* of tasks, i.e., the order among tasks does not matter, but the number of executions of each task does. For example, the states for $\langle t_1, t_2, t_3 \rangle$ and $\langle t_1, t_3, t_2 \rangle$ are the same, i.e., $[t_1, t_2, t_3]$. However, the states for $\langle t_1, t_2, t_9 \rangle$ and $\langle t_1, t_2, t_9, t_2 \rangle$ are not the same, i.e., $[t_1, t_2, t_9]$ and $[t_1, t_2^2, t_9]$ respectively, because the number of occurrences of each task matters.

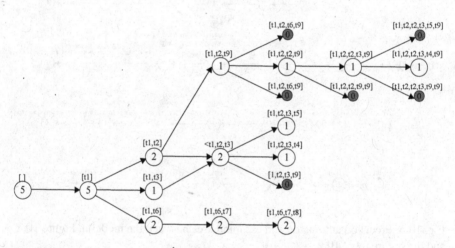

Fig. 10.6: Automaton from Λ^1 with multiset state representation for the running example medical process.

Figure 10.6 show the 1-alignment automaton for the medical process of Figure 9.2, considering the multiset state representation. This automaton contains differences with respect to its ordered homologous (cf. Figure 9.5). For example, instead of having two states $\langle t_1, t_2, t_3 \rangle$ and $\langle t_1, t_3, t_2 \rangle$ for prefixes $\langle t_1, t_2, t_3 \rangle$ and $\langle t_1, t_3, t_2 \rangle$, both prefixes are now represented as a single state $[t_1, t_2, t_3]$. This representation reduces the number of escaping arcs and hence increases precision values. Using multiset state representation and precision calculation as explained in Section 9.3, the model in Figure 10.4 and log in Table 10.2 used to motivate this section has a precision value of 1 (perfect). It is worth mentioning that in [14], the authors also propose the use of *set* as state representation. However, this is not applicable to our case: unlike sequence or multiset, a set does not preserve the number of occurrences of each task executed, and therefore, it may represent a (possible infinite) number of different model states. For example, given the model in Figure 9.2, the set $\{t_1, t_2, t_9\}$ represents $\langle t_1, t_2, t_9 \rangle, \langle t_1, t_2, t_9, t_2 \rangle, \langle t_1, t_2, t_9, t_2, t_9 \rangle, \ldots$.

10.3.2 Abstraction on the Direction

In the approach presented in Section 9.3, the prefixes of the complete tasks sequences are used to build the automaton. For example, given a complete task sequence $\langle t_1, t_2, t_3, t_4 \rangle$, the states constructed from the sequence are the empty sequence $\langle \rangle$ (corresponding with $\langle \bullet t_1, t_2, t_3, t_4 \rangle$, where $\bullet$ indicates a point of interest in the sequence), $\langle t_1 \rangle$ (for $\langle t_1 \bullet, t_2, t_3, t_4 \rangle$), $\langle t_1, t_2 \rangle$ (for $\langle t_1, t_2, t_3, t_4 \bullet t_3, t_4 \rangle$), $\langle t_1, t_2, t_3 \rangle$ (for $\langle t_1, t_2, t_3 \bullet t_4 \rangle$) and finally $\langle t_1, t_2, t_3, t_4 \rangle$ (for $\langle t_1, t_2, t_3, t_4 \bullet \rangle$). In other words, only the activities in the past are used and we move *forward* on the complete task sequences. This approach is used by all existing precision checking techniques [59, 60, 61].

In [14], the authors show that any point in the sequence (represented as $\bullet$) may represent two complementary visions: the *past* tasks seen until that point (as it has been shown above), but also the *future* tasks to come until the ending of the case. For instance, given $\langle t_1, t_2 \bullet t_3, t_4 \rangle$, $\langle t_1, t_2 \rangle$ are the tasks occurred, while $\langle t_3, t_4 \rangle$ are the tasks to happen. Both $\langle t_1, t_2 \rangle$ and $\langle t_3, t_4 \rangle$ are used in [14] as two different states that can be derived from the same point in the sequence. In this section, we use the same idea to present a *backward* precision measurement, that complements the *forward* approach presented before. The combination of metric results for both approaches will lead to a measurement unbiased by the direction of the precision checking. For the sake of clarity we will use ordered state representation to illustrate the remainder of the section, although the analogous procedure is applicable for unordered representation.

Let Λ be the option chosen to compute precision, i.e., Λ^1, Λ^* or Λ^R. In order to build the automaton for the backward precision measurement, we consider the prefixes of the *reversed* complete tasks sequences in $row_T(\Lambda)$. In other words, given $row_T(\gamma) = \langle t_1, t_2, t_3, t_4 \rangle$ of the alignment $\gamma \in \Lambda$, we use $row'_T(\gamma) = \langle t_4, t_3, t_2, t_1 \rangle$ to determine the states, resulting in the following 5 states: $\langle \rangle$ (corresponding with $\langle \bullet t_4, t_3, t_2, t_1 \rangle$), $\langle t_4 \rangle$ (for $\langle t_4 \bullet t_3, t_2, t_1 \rangle$), $\langle t_4, t_3 \rangle$ (for $\langle t_4, t_3 \bullet t_2, t_1 \rangle$), $\langle t_4, t_3, t_2 \rangle$ (for $\langle t_4, t_3, t_2 \bullet t_1 \rangle$) and finally $\langle t_4, t_3, t_2, t_1 \rangle$ (for $\langle t_4, t_3, t_2, t_1 \bullet \rangle$). Analogously, the set of complete tasks sequences of M is also reversed.[1] The rest of the precision checking is performed as it is described in Section 9.3.

Figure 10.7 shows an example of two automata for the trace $\langle a, b, c, d \rangle$, constructed by moving in forward direction (left) and by moving backward (right). No-

---

[1] Notice that, for the case of Petri nets with one unique initial and final markings, the set of all reversed complete transition sequences can be generated by the behavior of a net obtained from the original net by reversing its arcs and swapping their initial with final marking.

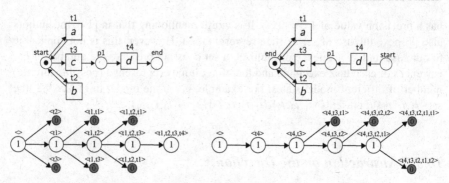

Fig. 10.7: Example of model and resulting automaton for the trace $\langle a,b,c,d \rangle$, with both forward and backwards approaches.

tice the difference of identified escaping arcs shown by the two automata. Finally, precision values obtained using forward and backward-constructed automaton can be combined (e.g., the average), resulting in a balanced precision metric unbiased by the direction of the automaton constructed. Note that more sophisticated and flexible combinations of both metrics are also possible. In Section 11.2, we investigate the differences in precision values produced by the various approaches using a variety of event logs and models.

10.4 Summary

This chapter provided alternatives to the basic alignment approach to analyze different types of scenarios when the fitness between observed and modeled behavior does not match perfectly. In particular, it explored the use of several optimal alignments and the directionality of those alignments.

Chapter 11
Handling non-Fitness in Practice

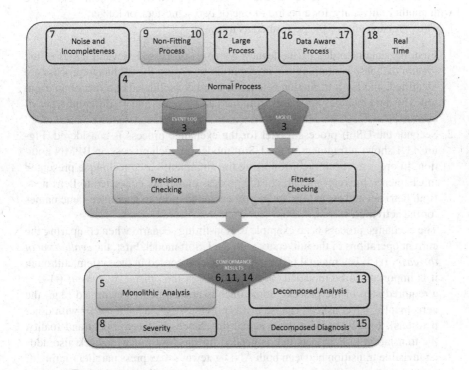

The previous chapter proposed a method to measure precision based on alignments for non-fitting cases. This chapter uses the University Case example to illustrate the applicability of the alignment-based approach to analyze possible precision anomalies in a non-fitting process. Additionally, the chapter provides an experimental analysis on the different characteristics of the approach. That concludes Part II

© Springer International Publishing AG 2016
J. Munoz-Gama: Conf. Check. ... in Process Mining, LNBIP 270, pp. 107–117, 2016.
DOI: 10.1007/978-3-319-49451-7_11

of the book focused on precision checking, while Part III will focus on fitness for large processes.

11.1 The University Case: The Exchange Process

This example was introduced in Chapter 1, where a university is considering to change its own system to buy a new BPM solution, USoft, to manage its processes. The university wants to analyze how good the USoft process models support its current processes. For that, a conformance analysis is proposed, where the models are analyzed with respect to the event logs recorded by the legacy system.

In this particular scenario, conformance checking is applied to analyze the precision of the USoft model and the legacy system event log, for the *exchange process*. The exchange process involves students who have applied for an academic exchange with another university, for a period of one or two semesters, or longer.

The precision analysis is performed as follows:

1. First, the data recorded by the legacy system is processed. Several tables of the system database containing the information regarding the exchange process are identified, and the information is extracted and consolidated in terms of an event log. The data includes 850 cases from the last 10 years, and 18 different types of events.

2. Second, the USoft process model for the exchange process is considered. Figure 11.1 shows a fragment of the USoft model for such process, in BPMN notation. In order to check precision using the alignment-based technique presented in Chapter 9, the relevant parts of the process model are converted to Petri nets. Both Petri net and event logs are pre-processed in order to guarantee same names for the activities.

3. The exchange process is an example of non-fitting scenario when comparing the current operations of the university with the USoft model. First, the *evaluation of the cases* (t15) is a manual task, and it is never recorded in the system, although it is important to be modeled. Second, although the *official agreement* (t14) is a required step of the process, sometimes is not explicitly performed (e.g., the agreement is never asked because it was previously included together with other petitions). Finally, besides these actual differences between model and reality, the translation to Petri nets that is needed for the conformance check also adds an invisible transition between both AND gateways to express parallel routing of control-flow; the event log does not contain events for this transition. Therefore, a precision checking based on alignments is selected to conduct the analysis.

4. Both modeled and observed behavior are aligned, and the result is used to perform the escaping arcs based precision checking. Figure 11.2 show a fragment of the alignments provided by the ProM plugin-in *Replay a Log in Petri Net for Conformance Analysis* by A. Adriansyah (top), and a fragment of the precision analysis performed over the alignments provided by the ProM plug-in *Precision*

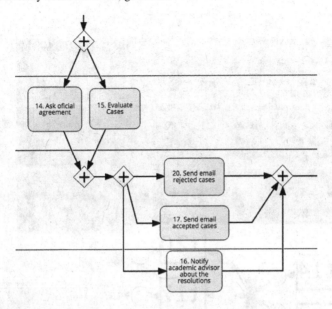

Fig. 11.1: Fragment of the exchange process model in USoft in BPMN notation.

for DPN, developed by F. Mannhardt. The results provided include a visualization of the Petri net coloring the places with precision problems (middle left), the minimal imprecise traces denoting the comparison between observed and possible arcs (middle right), and a summary of the precision and arcs (bottom).

5. After analyzing the results, the university concludes that the USoft model for the exchange process is precise enough for their requirements (0.75). The alignments confirm that t14 is sometimes skipped (not an important problem). However, a more serious precision problem is detected using the alignments: the model allows for the execution of t20, t17, and t16 in parallel. However, the log reflects that t16 must be performed before t17, and t20, resulting in escaping arcs. The university then confirms that this is a violation of the university guidelines, since the academic advisor needs to be notified before sending the emails.

6. Finally, the USoft exchange model is adapted to correct the precision problem identified in the analysis. Figure 11.3 shows a fragment of the adapted model.

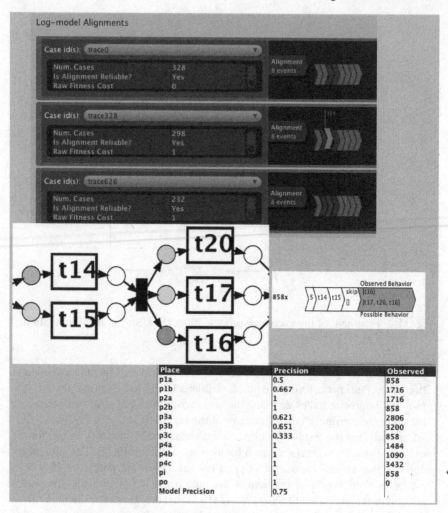

Fig. 11.2: Fragment of the precision results provided by *Replay a Log in Petri Net for Conformance Analysis* and *Precision for DPN* tools.

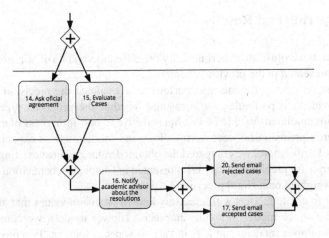

Fig. 11.3: Fragment of the adapted exchange process model in USoft in BPMN notation.

11.2 Experimental Results

In this section we illustrate experimentally the different variants of alignment-based precision presented in the previous chapters.

The first set of experiments was performed to evaluate the precision measurements provided. In particular, we measured whether the proposed precision approaches are unidimensional [87], i.e., not sensitive to non-fittingness of event logs. We measured precision between various logs and models whose expected values are known. Furthermore, we compared the obtained values against existing state-of-the-art metrics for precision: etc_P [59] (presented in Chapter 5), behavioral precision [88], and weighted behavioral precision [30].

In order to create models with precisely known precision values that are neither fully precise (trace model) nor fully imprecise (flower model), we combine precise models, flower models, and logs in various ways as follows. Two models, with disjoint set of activities, were combined by merging the end place of one with the initially marked place of another. The merged models were named according to the name of their original models, e.g., **PF** model is the result of merging the end place of completely precise **P** model with the initially marked place of the flower model **F**. Precision values were measured 30 times using 30 event logs, each consisting of 5,000 traces, generated by simulating the precise model (i.e., **PP**). For sake of completeness, we also measured the precision of the overfitting model (**P**) and the flower model (**F**) using 30 logs of 5,000 traces generated by simulating the **P** model. This way, *each log contains all the possible behavior of the model that generates it* (i.e., for any two activities that can follow each other directly according to the model, there is also a trace in the log where they follow each other directly).

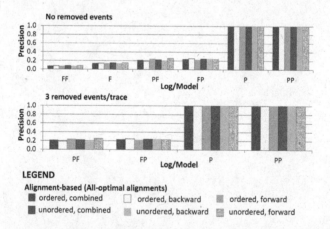

Fig. 11.4: Precision values of flower and over-precise logs/models and their combinations provided by alignment-based approach (i.e., computed using all optimal alignments, ordered, and forward-constructed automata). If all behavior are observed in the original logs, all measurements are insensitive to non-fitting traces.

The top part of Figure 11.4 shows the alignment-based precision values, measured using all optimal alignments per trace of the logs. The experiment with one and representative alignments per trace yields identical results. This result shows that by observing sufficiently enough behavior in the event logs, all alignment-based measures provide similar intuition about precision of models, i.e., overfitting models have high precision values and "flower" models have low precision values. Note that there are slight differences between various configurations of measures, i.e., states (ordered/unordered) and forward/backward constructed automata.

To evaluate the robustness of the measures against non-fitting logs, we took the models and logs from the previous experiments and created unfitting logs by removing n random events per trace from the fitting logs. Furthermore, the measurements are compared against existing measures. We use the CoBeFra tool [31] to measure behavioral precision [88] and weighted behavioral precision [30]) and use ProM 6 to measure etc_P. The bottom part of Figure 11.4, Figures 11.5 and 11.6 show some of the results.

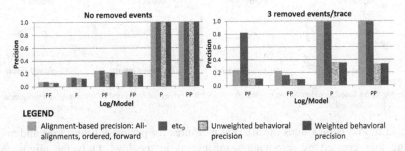

Fig. 11.5: Comparison between precision values obtained using alignment-based approach (i.e., computed using all optimal alignments, ordered, and forward-constructed automata) and other measures. Only the alignment-based approach is not sensitive to non-fitting logs/models.

The bottom part of Figure 11.4 shows that the measures proposed in this chapter are robust to fitness problems. Figure 11.5 shows a comparison between the precision values provided by alignment-based measures and other existing measures. For readability, we only show one alignment-based measure: the one computed using all-optimal alignments and forward-constructed automata whose states are constructed by taking into account activity ordering. Note that in cases where logs are perfectly fitting the models, all measures provide similar precision intuition. In fact, the alignment-based precision values shown in Figure 11.5 are the same as the etc_P values. However, in cases where logs are non-fitting, other measures may show misleading precision insights. The etc_P measure provides low precision for model **PF** with respect to perfectly fitting logs (i.e., 0.25). However, the value rises to 0.82 when 3 events are removed from the logs, because for all non-fitting traces, it ignores the rest of the traces after the first non-fitting event occur. Similarly, both weighted and unweighted behavioral precision measures provide lower precision values for non-fitting logs than the ones provided for perfectly fitting logs. Even for overly fitting models **P** and **PP**, both measures provide precision values below half (i.e., indicating the models are imprecise). This occurs because both measures mixed both perfectly-fitting and non-fitting traces in construction of artificial negative events, which leads to misleading construction of artificial negative events.

Figure 11.6 shows the influence of noise by removing some events in the logs. As shown in the figure, other than the alignment-based precision measure, precision values of all measures may change significantly even with only one event removed from all traces. Due to the randomness of the location of removed events, the etc_P measure may both increases or decreases in the presence of non-fitting traces. Both weighted and unweighted behavioral precision measures decreases when more events are removed because incorrect artificial negative events are introduced. Note that the number of negative events tends to decrease when traces in the log gets more diverse because of the removal of events.

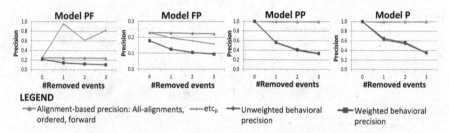

Fig. 11.6: Precision values of different measures for perfectly fitting logs and non-fitting logs created by removing some events in the logs. Only the alignment-based approach measure (i.e., computed using all optimal alignments, ordered, and forward-constructed automata) is insensitive to non-fitting logs.

The set of experiments also shows some interesting insights into the differences between alignment-based measures. Figure 11.7 shows a comparison between precision values provided by the two measures for models **PF** and **FP**. As shown in the figure, precision values of alignment-based measures provided by forward-constructed automata for model **PF** is higher than the values provided by backward-constructed automata for the same model, regardless of the noise level and the state representation (ordered/unordered). In contrast, the values provided by the latter is higher than the former for the **FP** model. This shows that the position of the precise part of the models influences precision values. Precision values are higher when the direction of constructed automata starts with precise part of process models. In this case, we clearly see the influence of forward/backward direction of constructed automata to precision values. To balance the influence, one of the simplest way is to take the average between the values provided by both directions. Figure 11.7 shows that the precision values obtained by combining both values are almost similar between model **PF** and **FP**.

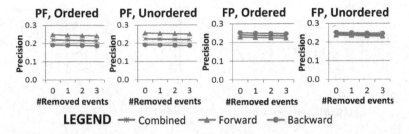

Fig. 11.7: Precision values of the **PF** and **FP** using all-alignments per trace, with different state representations (ordered/non-ordered) and direction (forward/backward). Higher precision is obtained when the direction of automata construction starts with precise part of the models.

Log	#Cases	#Events	Process Model		#Deviation/trace
			#Place	#Trans	
Bouw-1	139	3,364	33	34	9.75
Bouw-4	109	2,331	31	31	7.27
MLog1	3,181	20,491	15	12	5.33
MLog2	1,861	15,708	16	19	1.45
MLog3	10,271	85,548	24	21	14.50
MLog4	4,852	29,737	16	27	2.09
MLog5	25,846	141,755	14	24	1.21
IsalaLog	77	459	26	39	0.68

Table 11.1: Real-life logs and models used for experiments

To evaluate the applicability of the approach to handle real life logs, we used 8 pairs of process models and logs from two different domains (see Table 11.1), where 7 logs and models were obtained from municipalities in the Netherlands. In particular, we took the collections of logs and models from the CoSeLoG project [33]. The remaining pair of log and model is obtained from a hospital in the Netherlands[1]. The logs and models from municipalities are related to different types of building permission applications, while the hospital log is related to patient handling procedure. All processes have unlogged tasks, and some of the models allow loops. Table 11.1 shows an overview of the logs and models used in the experiments. #Deviations/trace column indicates the number of asynchronous moves after aligning all traces in the logs with their corresponding models. As shown in Table 11.1, all logs are not perfectly fitting to the corresponding models. We measure the precision values for all logs and the required computation time. The results are shown in Figure 11.8 and Figure 11.9.

Figure 11.8 reports precision values obtained for the real-life logs and models. Only the approach based on 1-alignment provides precision values for all real-life logs and models in the experiments. The approach based on all-optimal alignments per trace had out-of-memory problems when dealing with relatively complex process models and logs such as "Bouw-1" (33 places, 34 transitions), "Bouw-4" (31

---

[1] see http://www.healthcare-analytics-process-mining.org/

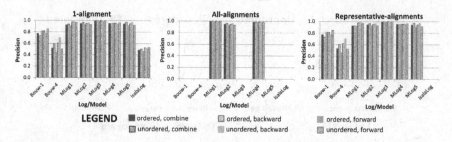

Fig. 11.8: Precision values of real-life logs and models. Only the 1-alignment approach manages to provide precision results for all logs/models.

places, 31 transitions), and "MLog-3" (24 places, 21 transitions). Precision measurements based on representative of optimal alignments also had the same problems dealing with the hospital log (i.e., "IsalaLog"). Although the model of the log is relatively small, it contains many unlogged tasks (tasks whose execution are not logged), allows loops, and allows many interleaving activities such that the size of state space required to compute even the representative of all optimal alignments is large and does not fit memory.

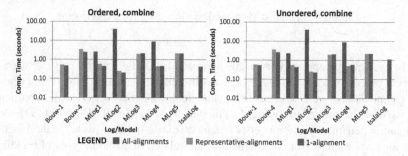

Fig. 11.9: Computation time comparison of alignment-based precision measurement using combined values (from backward and forward automata construction). Y-axis values are shown in a *logarithmic* scale.

Nevertheless, notice the similarity of the computed precision values using all three alignments (1-align, all-align, and representatives). From all pairs of logs and models, only 2 of them have precision value below 0.7. This shows that in reality, process models are made to be relatively precise such that meaningful insights into the process can be obtained. Interestingly, different precision values are provided by different measures in the experiment with log and model "Bouw-4" when both one and representative alignments are used. The precision value provided by ordered-forward measure for the model is around 0.44 (showing imprecision) while the unordered-backward precision measure provides a value of 0.7 (i.e., precise).

This indicates that more observations are required to measure the particular log and model accurately.

Figure 11.9 reports the computation time required to measure precision of real-life logs and models using alignment-based approach with combined precision values between forward and backward-constructed automata. The y-axis of the charts are shown in logarithmic scale. As shown in the figure, the computation time of precision measurement with all-alignments takes much longer than the ones required by one or representative alignments. All measurements using 1-alignment/representative alignments were computed in less than 10 seconds. Notice the similarity between the left and right graph on the figure (except the IsalaLog that has out-of-memory problem in the approach with representative alignments). In fact, we obtained identical results for all other combination of state representations (ordered/unordered) and directions where automata is constructed (forward/backward). This shows that the different directions of the automata construction and state representations are not significantly influencing computation time. Instead, most computation time of precision measurement is spent in the alignment of logs and process models. Another interesting observation is that the time spent to compute representative alignments are similar to the time spent to compute 1-alignment. Thus, we recorded the number of generated representatives for the experiments and other statistics to investigate this.

In conclusion, the experiments show that the use of alignments provides a precision measurement that works well on real-life logs, with reasonable computation time and memory requirements, and reliable results. Moreover, the 1-alignment variant is preferred when time is a constraint or when only an overview analysis is required, while the all-alignment variant provides a more reliable result balancing between possible optimal alignments.

Part III
Decomposition in Conformance Checking

Chapter 12
Decomposing Conformance Checking

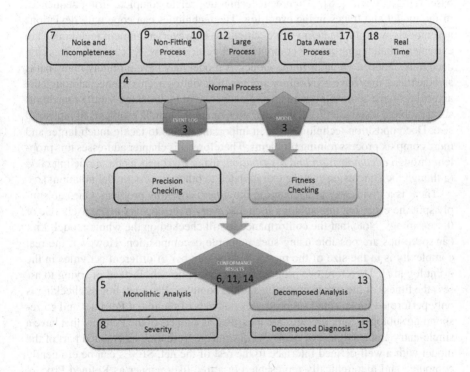

Conformance checking is a time-consuming analysis, and the diagnose of conformance problems is not free of difficulties. This chapter provides a general decomposition of the original problem in parts, alleviating its computation time, and providing mechanism to locally understand the detected problems. In later chapters, we will apply this idea to analyze the fitness of large processes.

© Springer International Publishing AG 2016
J. Munoz-Gama: Conf. Check. ... in Process Mining, LNBIP 270, pp. 121–127, 2016.
DOI: 10.1007/978-3-319-49451-7_12

12.1 Introduction

While in the previous chapters we analyzed the precision dimension, in these part we focus on the fitness dimension of conformance checking. In real-life situations, event logs often do not fit its corresponding models, i.e., some log traces cannot be fully reproduced in the model. These non-fitting situations should be communicated to the stakeholders, in order to take decisions on the process object of study. However, in reality process models can be non-deterministic, which complicates the analysis. Non-determinism may arise when the model contains silent or duplicate activities, which is often the case in practice. Moreover, the presence of noise in the log – e.g., incorrectly logged events or abnormal behavior – complicates even more the algorithmic detection of non-fitting situations. Due to this, the initial fitness approaches based on replaying log traces in a model in order to assess whether a trace can fit a model [77] have been recently reconsidered, giving rise to approaches based on alignment [18]. Alignment techniques relate complete firing sequences of the model and traces in the event log. The techniques can cope with deviations and models with duplicate/invisible activities. However, alignment techniques are extremely challenging from a computational point of view. Traces in the event log need to be mapped on paths in the model. A model may have infinitely many paths and the traces may have an arbitrary amount of deviating events. Hence, although the algorithms have demonstrated to be of great value for undertaking small or medium-sized problem instances [7, 2], they are often unable to handle problems of industrial size. Decomposition techniques are an important means to tackle much larger and more complex process mining problems. Therefore, this chapter addresses this problem through *decomposition and distribution*, using Petri nets as the modeling class (although the conclusions can be extrapolated to other process model notations).

There is a trivial way to decompose the fitness checking problem. One can simply split the event log into sublogs such that every trace appears in precisely one of these sublogs. Note that the conformance is still checked on the whole model. Linear speed-ups are possible using such a simple decomposition. However, the real complexity is in the size of the model and the number of different activities in the event log [17]. Therefore, we propose a different approach. Instead of trying to assess the fitness of the whole event log and the complete Petri net, fitness checking is only performed for selected subprocesses (subnets of the initial Petri net and corresponding sublogs). Subprocesses are identified as subnets of the Petri net that have a single-entry and a single-exit node (*SESE*), thus representing an isolated part of the model with a well-defined interface to the rest of the net. SESEs can be efficiently computed and hierarchically represented in a tree-like manner as Refined Process Structured Tree (RPST) [70].

12.2 Single-Entry Single-Exit and Refined Process Structure Tree

The intuitive idea behind the decomposition technique in this chapter is to find subgraphs that have a simple interface with respect to the rest of the net. The following set of definitions formalizes the idea of *Single-Entry Single-Exit* subnet and the corresponding decomposition. The underlying theory dates back to the seminal work of Hopcroft and Tarjan in the seventies [47], but recent studies have made considerable progress into making the algorithms practical when applied to process models [70, 69]. We start by defining the graph structure used for decomposing a process model: the *workflow graph*. Given a workflow net[1], the workflow graph represents the arcs between the nodes of the net.

Definition 12.1 (Workflow Graph) *Let $WN = (P, T, F, start, end)$ be a workflow net, where P are the places, T are the transitions, and F are the arcs. We define the* workflow graph *of WN simply as the directed graph $G(WN) = (N, F)$ where no distinctions are made between places and transitions, i.e., $N = P \cup T$ represent the nodes of the graph, and F the arcs. We can omit the parameter WN whenever the context is clear.*

An example of workflow graph Figure 12.1b (only the graph but not the squares) is shown for the workflow net in Figure 12.1a.

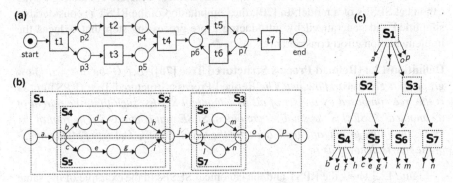

Fig. 12.1: An example of (a) workflow net, (b) its workflow graph, and the (c) RPST with its SESE decomposition.

In the remainder, the following context is assumed: Let G be the workflow graph of a given WN, and let $G_S = (V_S, S)$ be a connected subgraph of G formed by a set of edges S and the vertices $V_S = \Pi(S)$ induced by S.[2]

---

[1] Although the approach presented in this chapter can be generalized to general Petri nets with several sources and sinks, for the sake of clarity in this paper we restrict to the workflow case with only one source and only one sink [70].

[2] $\Pi(R) = \bigcup_{(a,b) \in R} \{a, b\}$ is the set of elements referred to by relation $X \subseteq A \times B$.

Definition 12.2 (Subnet nodes [70]) *A node $x \in V_S$ is interior with respect to G_S iff it is connected only to nodes in V_S; otherwise x is a boundary node of G_S. A boundary node y of G_S is an entry of G_S iff no incoming edge of y belongs to S or if all outgoing edges of y belong to S. A boundary node y of G_S is an exit of G_S iff no outgoing edge of y belongs to S or if all incoming edges of y belong to S.*

For example, let us consider the arcs $S_4 = \{b,d,f,h\}$ of Figure 12.1, and the set of vertices induced by them $V_{S_4} = \{t_1, p_2, t_2, p_4, t_4\}$. t_1 and t_4 are boundary nodes, while p_2, t_2, p_4 are interior. Moreover, t_1 is an entry, while t_4 is an exit.

As the next definition formalizes, a SESE is a special type of subgraph with a very restricted interface with respect to the rest of the graph:

Definition 12.3 (Single-Exit-Single-Entry [70]) *A set of edges $S \subseteq F$ is a SESE (Single-Exit-Single-Entry) of graph $G = (N, F)$ iff G_S has exactly two boundary nodes: one entry and one exit. A SESE is trivial if it is composed of a single edge. S is a canonical SESE of G if it does not partially overlap with any other SESE of G, i.e., given any other SESE S' of G, they are nested ($S \subseteq S'$ or $S' \subseteq S$) or they are disjoint ($S \cap S' = \emptyset$). By definition, the source start of a workflow net is an entry to every fragment it belongs to and the sink end of the net is an exit from every fragment it belongs to.*

The decomposition based on canonical SESEs is a well studied problem in the literature, and can be computed in linear time. In [86], the authors proposed the algorithm for constructing the RPST, i.e., a hierarchical structure containing all the canonical SESEs of a model. In [70], the computation of the RPST is considerably simplified and generalized by introducing a pre-processing step that reduces the implementation effort considerably.

Definition 12.4 (Refined Process Structured Tree [70]) *Let G be the workflow graph of a given workflow net. The* Refined Process Structured Tree (RPST) *of G is the tree composed by the set of all its canonical SESEs, such that, the parent of a canonical SESE S is the smallest canonical SESE that contains S. The root of the tree is the entire graph, and the leaves are the trivial SESEs. The set of all the nodes of the tree is denoted as $\mathbb{S}$.*

Figure 12.1 show the RPST and the canonical SESEs in the example of the same figure. In the remainder of the chapter, we will refer to canonical SESEs resulting from the RPST decomposition simply as SESEs. Also note that the SESEs are defined as a set of edges (i.e., S) over the workflow graph (not as subgraphs, i.e., G_S). However, for simplicity and when the context is clear, we will use the term SESE to refer also to the subgraph induced by those edges. We will denote as $PN^S = (P^S, T^S, F^S)$ the Petri net determined by the SESE S, i.e., $PN^S = (P \cap \Pi(S), T \cap \Pi(S), F \cap S)$. The nodes (either transitions or places) determined by S are denoted as N^S, i.e., $(P \cup T) \cap \Pi(S)$.

12.3 Decomposing Conformance Checking using SESEs

It is well known that checking conformance of large logs and models is a challenging problem. The size of log and model and the complexity of the underlying process strongly influence the time needed to compute fitness and to create optimal alignments. *Divide-and-conquer* strategies are a way to address this problem [64, 63, 62, 9]. As indicated before, we do not just want to partition the traces in the event log (providing a trivial way to distribute conformance checking). The potential gains are much higher if the model is also decomposed and traces are split into smaller ones. To decompose conformance checking problems, the overall system net *SN* is broken down into a collection of subnets $\{SN^1, SN^2, \ldots, SN^n\}$ such that the union of these subnets yields the original system net.

Definition 12.5 (Decomposition) *Let $SN = (SNM_I, M_F)$ be a system net where $WN = (P, T, F, start, end)$. $D = \{SN^1, SN^2, \ldots SN^n\}$ is a decomposition of SN if and only if:*

- $P = \bigcup_{1 \leq i \leq n} P^i$,
- $T = \bigcup_{1 \leq i \leq n} T^i$,
- $F = \bigcup_{1 \leq i \leq n} F^i$ where $F^i \cap F^j = \emptyset$ for $1 \leq i < j \leq n$.

Note that each place or transition can be shared among different subnets, while each arc resides in just one subnet.

Any decomposition that satisfies Definition 12.5 may be considered for decomposing a conformance problem, e.g., subnets containing only one arc, or subnets randomly grouping distant nodes on the net. However, given that the ultimate goal of a decomposition is to be able to diagnose, comprehend and understand conformance problems, the use of meaningful decompositions is preferred, i.e., *SESEs*. Given the structure of a SESE where a unique single entry and a unique single exit exist, a SESE becomes an appropriate unit of decomposition. Intuitively, each SESE may represent a subprocess within the main process (i.e., the interior nodes are not connected with the rest of the net), and the analysis of every SESE can be performed independently. The *RPST* of a net can then be used to select a possible set of SESEs forming a decomposition. As it shown in Proposition. 12.6, any *transverse cut* over the RPST defines a decomposition.

Proposition 12.6 (SESE decomposition) *Let $SN = (SNM_I, M_F)$ be the system net of the workflow net $WN = (P, T, F, start, end)$. Consider the RPST decomposition of WN, where $\mathbb{S}$ represents all the SESEs in the RPST. We define a transverse-cut over the RPST as a set of SESEs $\mathbb{D} \subseteq \mathbb{S}$ such that any path from the root to a leaf of RPST contains one and only one SESE in $\mathbb{D}$. Given a transverse-cut $\mathbb{D} = \{S_1, S_2, \ldots S_n\}$, let the decomposition $D_{\mathbb{D}}$ be defined as $D_{\mathbb{D}} = \{SN^{S_1}, SN^{S_2}, \ldots SN^{S_n}\}$, where $SN^{S_i} = (PN^{S_i}, M_I \upharpoonright_{P^{S_i}}, M_F \upharpoonright_{P^{S_i}})$, i.e., the Petri net determined by the SESE S_i, and the projection of the initial and final markings on the places of the subnet. The decomposition $D_{\mathbb{D}}$ derived from the SESEs satisfies the definition of decomposition given in Definition 12.5*

Proof. By definition of RPST, the arcs of each SESE in the RPST are contained in one, and only one, of its children (unless it is a trivial SESE). Therefore, any transverse-cut set of SESEs contains all the arcs, where each arc only appears in only one SESE.

Proposition 12.7 (A SESE decomposition from RPST exists) *Given any RPST, a decomposition always exists.*

Proof. Given any RPST, the root (i.e., the whole net) defines a decomposition. In addition, the set of all the leaves (i.e., the trivial SESEs with only one arc) also defines a decomposition.

As it is claimed in Proposition 12.7, the overall net is, by definition, a decomposition by itself. But it is obvious to see that this trivial way of decomposition does not alleviate the initial conformance problem. On the other hand, a decomposition formed only by trivial SESEs will produce meaningless components, and at the same time, the posterior analysis will have to deal with the analysis overhead produced by the creation of the numerous components. A decomposition which lays in between the aforementioned extremes seems more interesting from the practical point of view, i.e., to generate components large enough to become meaningful subprocesses, but whose size can be handled in practice. Hence, the algorithm proposed in Algorithm 12.1 can generate a decomposition which limits the maximum size of each component to k in order to control the size and complexity of individual components.

Algorithm 12.1 k-decomposition algorithm

---

procedure k-DEC(RPST,k)
   $V = \{root(RPST)\}$
   $D = \emptyset$
   while $V \neq \emptyset$ **do**
      $v \leftarrow pop(V)$
      if $|v.arcs()| \leq k$ **then** $D = D \cup \{v\}$
      else $V = V \cup \{children(v)\}$
   return D

Algorithm 12.1 shows how to compute a k-decomposition, for any k such that $1 \leq k \leq |F|$ (where $|F|$ stands for the number of arcs of the overall net). The algorithm keeps a set of nodes that conform the decomposition (D) and a set of nodes to consider (V). Initially V contains the root of the RPST, i.e., the overall net. Then, the algorithm checks, for each node v to consider, if v satisfies the k property, i.e., the number of arcs of SESE v is less or equal than k. If this is the case, v is included in the decomposition. If not, it discards v and includes the RPST children of v into the nodes to consider. Note that, given any RPST, a k-decomposition always exists, i.e., in worst case, the decomposition formed by all the leaves of the RPST will satisfy the definition. The algorithm proposed has linear complexity with respect to the size of the RPST, and termination is guaranteed by the fact that the size of the component is reduced in every iteration.

12.4 Summary

This chapter presented the decomposition of the conformance checking problem, i.e., the process model is decomposed in subprocesses with a single entry and a single exit node, and they are analyzed independently. This decomposition reduces the computation time, and makes it possible to understand the conformance problems in local way. However, in order to guarantee certain conformance properties over the general case, the decomposition must satisfy certain characteristics. In the next chapter, the SESE decomposition is extended to fulfill such characteristics.

Chapter 13
Decomposing for Fitness Checking

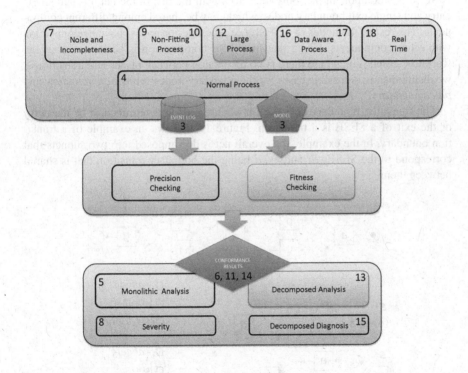

In the previous chapter, a decomposing method based on single entry single exit components was presented to generically decompose the conformance problem. This chapter presents an extension of the method to provided a so-called valid decomposition, i.e., a decomposition that preserves certain conformance guarantees over the general problem, even when it is analyzed in a decomposed manner. In the next chapter, the fitness analysis based on decomposition is put into practice.

© Springer International Publishing AG 2016
J. Munoz-Gama: Conf. Check. ... in Process Mining, LNBIP 270, pp. 129–139, 2016.
DOI: 10.1007/978-3-319-49451-7_13

13.1 Introduction

Decomposing conformance analysis is helpful in the diagnosis and understanding of the conformance problems. The model is divided into parts and conformance checking is performed over each part individually. However, in order to guarantee the relation between the local analysis and the general problem, the decomposition applied must satisfy certain properties, i.e., it must be a valid decomposition. The remaining of this chapter adapts the SESE decomposition presented in the previous chapter, in order to provide a valid decomposition.

13.2 Bridging a Valid Decomposition

A SESE is a component that only interfaces with the rest of the net through single entry and single exit boundary nodes, which may be shared among different components. The rest of the nodes of a SESE (i.e., the interior nodes) have no connection with other components. Given that the SESE computation is performed over the workflow graph (i.e., where there is no distinction between places and transitions), we distinguish two possible cases for the boundary nodes: *transition boundary* and *place boundary*.

The transition boundary case occurs when the node determined to be the entry or the exit of a SESE is a transition. Figure 13.1 shows an example of a transition boundary. In the example, the overall net is decomposed into two subnets that correspond to the SESEs S_1 and S_2, d being the boundary transition that is shared between them.

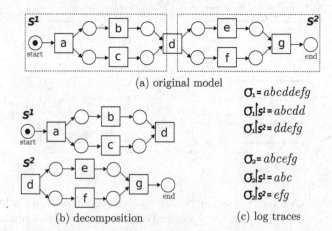

(a) original model

$\sigma_1 = abcddefg$

$\sigma_1|_{S^1} = abcdd$

$\sigma_1|_{S^2} = ddefg$

$\sigma_2 = abcefg$

$\sigma_2|_{S^1} = abc$

$\sigma_2|_{S^2} = efg$

(b) decomposition   (c) log traces

Fig. 13.1: Example of decomposition with transition boundary.

As it is proven in [9], a *transition boundary* decomposition represents no problem from a conformance point of view, i.e., given a decomposition with only transition boundaries, a log trace fits the overall net if and only if it fits all the subnets. The reason for that is that when a transition is shared between subnets, the label of the transition is used to synchronize the subnets that contain that transition on their boundaries, ensuring that the models' decisions to reproduce that label are done jointly. Consider the decomposition $D_\mathbb{D} = \{SN^{S_1}, SN^{S_2}\}$ from the example of Figure 13.1, where $SN^{S_1} = (PN^{S_1}, [start], [])$ and $SN^{S_2} = (PN^{S_2}, [], [end])$ are the systems nets derived from the SESEs S_1 and S_2. Consider the trace $\sigma_1 = abcddefg$ shown in Figure 13.1c. Such trace does not fit the overall net due to the double d. The projection of that trace on SN^{S_1} and SN^{S_2} results in $\sigma_1 \lceil_{T} s_1 = abcdd$ and $\sigma_1 \lceil_{T} s_2 = ddefg$ respectively (cf. Figure 13.1c). Note that, although $\sigma_1 \lceil_{T} s_2$ fits SN^{S_2} (on SN^{S_2}, the preset of d is empty hence it can fire more than once), $\sigma_1 \lceil_{T} s_1$ does not fit SN^{S_1}. Hence, the trace σ_1 that does not fit the overall net, does not fit all the subnets (at least there is one that is not fitting). A similar situation happens with the trace $\sigma_2 = abcefg$ (where no d appears), i.e., trace σ_2 does not fit the overall net, hence $\sigma_2 \lceil_{T} s_1$ does not fit SN^{S_1} or $\sigma_2 \lceil_{T} s_2$ does not fit SN^{S_2}. In the latter example, actually both do not fit.

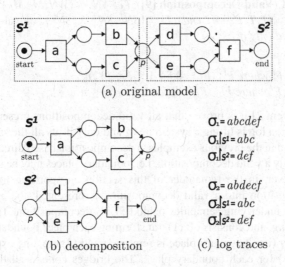

(a) original model

(b) decomposition   (c) log traces

Fig. 13.2: Example of decomposition with place boundary.

On the other hand, the case of *place boundary* is different. When the boundary (entry or exit) is a place, it is shared between two or more subnets. However, the arcs connected to the place (the ones in charge of producing and consuming tokens) are split amongst the subnets. This makes the place unable to synchronize, and therefore, it is impossible to analyze the different subnets in isolation. The example in Fig. 13.2 reflects this situation. The original net is decomposed into two subnets, $D_\mathbb{D} = \{SN^{S_1}, SN^{S_2}\}$, corresponding with the SESEs S_1 and S_2, and being p

the boundary place shared by both subnets. It can be seen that the arcs that produce tokens in p and the ones that consume tokens from p are distributed into different subnets. Consider now the log traces $\sigma_1 = abcdef$ and $\sigma_2 = abdecf$ of Fig. 13.2. While σ_1 fits the overall net, σ_2 does not. However, the projections of both traces on T^{S_1} and T^{S_2} are the same (cf. Fig. 13.2). This problem materializes when we analyze the subnets. Firstly, given that any arc that produces tokens in p is contained in PN^{S_1}, we need to consider an initial marking for SN^{S_2} different than $[]$ (otherwise, the subnet would be deadlocked initially). If we consider the initial marking $[p]$, $\sigma_1 \lceil_{T^{S_2}}$ does not fits SN^{S_2}. Therefore the fitness correctness is not preserved, i.e., a trace that fits the overall net like σ_1 must fit all the subnets. On the other hand, if we consider the initial marking with two (or more) tokens on p (i.e., $[p^2]$), $\sigma_2 \lceil_{T^{S_2}}$ fits SN^{S_2} (similarly, $\sigma_2 \lceil_{T^{S_1}}$ fits SN^{S_1}). However σ_2 is a non-fitting trace of the overall net, and consequently, it must not fit all the subnets. Therefore, when the decomposition contains place boundaries, the preservation of the fitness correctness is not guaranteed.

In [9] the definition of decomposition is revisited to propose the so called *valid decomposition*, i.e., a decomposition that only shares transitions (but not places nor arcs).

Definition 13.1 (Valid Decomposition [9]) *Let* $SN = (WN, M_I, M_F)$ *be a system net where* $WN = (P, T, F, start, end)$. $D = \{SN^1, SN^2, \ldots SN^n\}$ *is a valid decomposition of SN if and only if:*

- $T = \bigcup_{1 \leq i \leq n} T^i$,
- $P = \bigcup_{1 \leq i \leq n} P^i$ *where* $P^i \cap P^j = \emptyset$ *for* $1 \leq i < j \leq n$,
- $F = \bigcup_{1 \leq i \leq n} F^i$ *where* $F^i \cap F^j = \emptyset$ *for* $1 \leq i < j \leq n$.

In [9, Theorem 2] it is proven that all valid decompositions preserve the fitting correctness, i.e., a log is fitting a system net if and only if fits all the subnets.

As illustrated in the previous examples, a decomposition based directly on SESEs is not necessarily a valid decomposition, i.e., boundary places may be shared among subnets. However, in the remainder of this section an approach to transform a SESE decomposition into a valid decomposition is presented; the approach tries to preserve the underlying semantics behind SESE decomposition. This technique is called *bridging*, and consists of: (1) transforming each place boundary into a transition boundary (i.e., boundary place is removed) and (2) creating explicit subnets (called *bridges*) for each boundary place. The bridges contain all the transitions connected with the boundary place, and they are in charge of keeping the place synchronized among subnets. In addition, the boundary places together with the arcs connected to them are removed from the original subnets. Formally:

Definition 13.2 (Bridging a SESE decomposition) *Let* $\mathbb{D} = \{S_1, \ldots S_n\}$ *be the SESE decomposition of the workflow net* $WN = (P, T, F, start, end)$. *Let* $I_{\mathbb{D}} = \{i_1, \ldots, i_n\}$ *and* $O_{\mathbb{D}} = \{o_1, \ldots, o_n\}$ *be the set of all entry and exit nodes of the SESEs in* $\mathbb{D}$. $B = \{p_1, \ldots, p_k\} = ((I_{\mathbb{P}} \cup O_{\mathbb{P}}) \cap P) \setminus \{start, end\} = (I_{\mathbb{P}} \cap O_{\mathbb{P}}) \cap P$ *is the set of boundary places, i.e., entry and exit nodes of the SESEs that are places but not the source*

or sink place of the workflow net WN. The decomposition after applying bridging
$\mathbb{D}' = \{S'_1, \ldots S'_n, B_1 \ldots B_k\}$ *of* $\mathbb{D}$ *is constructed as follows:*

- *For all* $1 \leq i \leq n$: $S'_i = \{(x,y) \in S_i \mid \{x,y\} \cap B = \emptyset\}$ *(boundary places are removed from the SESEs).*
- *For* $1 \leq j \leq k$: $B_j = \{(x,y) \in A \mid p_j \in \{x,y\}\}$ *(bridges are added).*

$D_{\mathbb{D}'} = \{SN^{S'_1}, \ldots SN^{S'_n}, SN^{B_1} \ldots SN^{B_k}\}$ *represents the decomposition constructed from* $\mathbb{D}'$.

Figure 13.3 illustrates the effects of the bridging on the example previously shown in Fig. 13.2. In this case, the boundary place p (and its arcs) are removed from S_1 and S_2, and a bridge B_1 is created. Note that now, the transitions connected to p (i.e., b, c, d and e) are shared (instead of p), keeping the synchronization among components, and making $D_{\mathbb{D}'}$ a valid decomposition.

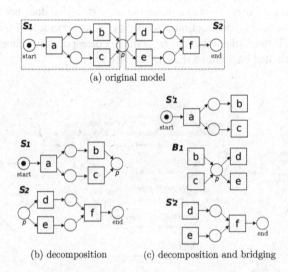

(a) original model

(b) decomposition   (c) decomposition and bridging

Fig. 13.3: Example of decomposition with bridging.

Proposition 13.3 shows that the decomposition derived from applying SESE decomposition and then bridging results in a valid decomposition, according to Def. 13.1.

Proposition 13.3 (Bridging results in valid decomposition) *Let* $\mathbb{D}' = \{S'_1, \ldots S'_n, B_1 \ldots B_k\}$ *be obtained from a SESE decomposition after applying bridging. The decomposition* $D_{\mathbb{D}'} = \{SN^{S'_1}, \ldots SN^{S'_n}, SN^{B_1} \ldots SN^{B_k}\}$ *is a valid decomposition according to Def. 13.1.*

Proof. By construction, a SESE decomposition only shares transitions and places. After applying the bridging, all the shared places are removed, creating explicit components with only one instance of these places.

Moreover, given that the bridging produces a valid decomposition, it also preserves the property that a trace in the log fits the overall process model if and only if each subtrace fits the corresponding process fragment. Hence, fitness checking can be decomposed using SESEs and bridges.

Proposition 13.4 (SESE-based Fitness Checking can be decomposed) *Let L be a log and $SN = (WN, M_I, M_F)$ be a system net where WN is a workflow net. Let $D_{\mathbb{D}'} = \{SN^1, SN^2, \dots SN^n\}$ be a valid decomposition resulting of the application of the SESE decomposition and bridging over WN. Let $SN^i = (PN^i, M_I^i, M_F^i)$, where $PN^i = (P^i, T^i, A^i)$.*

A trace $\sigma \in L$ fits SN (i.e., $(WN, M_I)[\sigma\rangle(WN, M_F)$) if and only if it fits all the parts, i.e., for all $SN^i \in D_{\mathbb{D}'}$, $(PN^i, M_I^i)[\sigma\restriction_{T^i}\rangle(PN^i, M_F^i)$.

Proof. Special case of the more general Theorem 2 in [9]. If the overall trace σ fits SN, then each of the projected traces $\sigma\restriction_{T^i}$ fits the corresponding subnet. If this is not the case, then at least there exist one projected trace $\sigma\restriction_{T^i}$ that does not fit. But this is impossible because, by construction, each subnet is a relaxation of the behavior of the overall net. If the projected traces $\sigma\restriction_{T^i}$ fit the corresponding subnets, then these traces can be stitched back into a trace σ that fits SN.

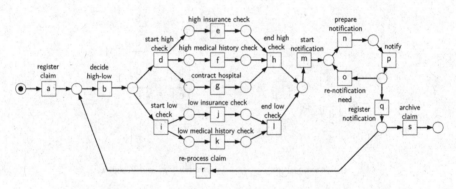

Fig. 13.4: Running example: claims in a insurance company.

Let us consider the example of Figure 13.5 to illustrate the decomposition of the conformance checking proposed in this Chapter. The model in figure was inspired by a similar model presented in [26] and represents the possible situations to handle claims in a insurance company.

Figure 13.5 shows the SESE decomposition by a size k of 15. Let us show how the fitness problems are now identified in a decomposed manner. For that, we will use the trace $\sigma = abijlmnpqnpqs$. Given σ and the SESEs provided in Figure 13.5, the only ones that reveal fitness anomalies are S_3', B_4 and B_6 (for the other components we can find perfect alignments when projecting σ to the activities of the component). The alignment for S_3' is:

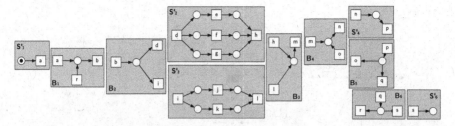

Fig. 13.5: Components resulting from 15-decomposition and bridging for the running example of Figure 13.4.

$$\begin{array}{|c|c|c|c|} \hline i & j & \gg & l \\ \hline i & j & k & l \\ \hline \end{array}$$

which reveals that the mandatory check of the medical history is missing in the log. Analogously, the alignment for B_4 is:

$$\begin{array}{|c|c|c|c|} \hline m & n & \gg & n \\ \hline m & n & o & n \\ \hline \end{array}$$

which identifies the explicit need to notify the client again, an action missing in the log but required by the model. Finally, the alignment for B_6:

$$\begin{array}{|c|c|c|} \hline q & q & s \\ \hline \gg & q & s \\ \hline \end{array}$$

reveals another fitness problem for trace σ: the system has stored in the log an early registration of the notification which was not meant at that point in time, since notifications were later sent and according to the model, the registration is only expected to be done at the end of the case.

By Prop. 13.4 and the fitness problems identified at some of the components in the decomposition, we can infer that the whole model does not fit the log.

13.3 Decomposition with invisible/duplicates

So far, the approach presented in this chapter assumes that all the Petri net transitions are associated with a unique single activity, i.e., a transition could be unambiguously identified by its label. In this section we lift this assumption in order to consider invisible and duplicate transitions. An invisible transition is a transition without an associated activity, e.g., transitions included for routing purposes. Duplicate transitions are transitions with the same activity associated. For example, consider the net of Figure 13.6, which is a slight variation of the example in Figure 13.4. This model contains an invisible transition (represented in black) which is used to *skip* the execution of *contract hospital*, i.e., now *contract hospi-*

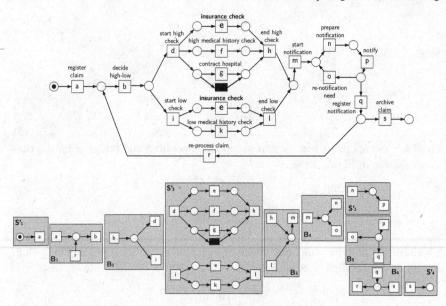

Fig. 13.6: Variant of the running example of Figure 13.4 including invisible and duplicates (top), and its corresponding decomposition (bottom).

tal is optional. Moreover, the new model does not distinguish between *high insurance check* and *low high insurance check*, but the same action *insurance check* is modeled in two different parts of the model, i.e., is a duplicate activity. The Petri net definition is extended considering now a labeling function $l \in T \not\rightarrow \mathscr{U}_A$ where $\mathscr{U}_A$ is some universe of activity labels. Additionally, if a transition $t \notin dom(l)$, it is called invisible. $T_v(PN) = dom(l)$ is the set of *visible* transitions in *PN*. $T_v^u(PN) = \{t \in T_v(PN) \mid \forall_{t' \in T_v(PN)}\ l(t) = l(t') \Rightarrow t = t'\}$ is the set of *unique* visible transitions in *PN* (i.e., there are no other transitions having the same visible label)

As it has been illustrated previously in this chapter, when a net is decomposed, the labels of the transitions are used to synchronize and preserve the fitness properties. However, sharing invisible and duplicate transitions among subnets generates ambiguity invalidating this synchronization. Thus, the definition of valid decomposition presented in Definition 13.1 is refined to consider invisible and duplicates, i.e., only unique visible transitions can be shared among subnets.

Definition 13.5 (Valid Decomposition with Invisible and Duplicates[9]) *Let* $SN = (WN, M_I, M_F)$ *be a system net where* $WN = (P, T, F, l, start, end)$. $D = \{SN^1, SN^2, \ldots SN^n\}$ *is a* valid *decomposition of SN if and only if:*

- $SN^i = (PN^i, M_I^i, M_F^i)$ *is a system net with* $PN^i = (P^i, T^i, F^i, l^i)$ *for all* $1 \le i \le n$,
- $l^i = l\!\restriction_{T^i}$ *for all* $1 \le i \le n$,
- $P^i \cap P^j = \emptyset$ *for* $1 \le i < j \le n$,
- $T^i \cap T^j \subseteq T_v^u(SN)$ *for* $1 \le i < j \le n$, *and*
- $SN = \bigcup_{1 \le i \le n} SN^i$.

Let $SN = (WN, M_I, M_F)$ with $WN = (P, T, F, l, start, end)$ be a system net with valid decomposition $D = \{SN^1, SN^2, \ldots, SN^n\}$. We can observe the following properties:

- each place appears in precisely one of the subnets, i.e., for any $p \in P$: $|\{1 \leq i \leq n \mid p \in P^i\}| = 1$,
- each invisible transition appears in precisely one of the subnets, i.e., for any $t \in T \setminus T_v(SN)$: $|\{1 \leq i \leq n \mid t \in T^i\}| = 1$,
- visible transitions that do not have a unique label (i.e., there are multiple transitions with the same label) appear in precisely one of the subnets, i.e., for any $t \in T_v(SN) \setminus T_v^u(SN)$: $|\{1 \leq i \leq n \mid t \in T^i\}| = 1$,
- visible transitions having a unique label may appear in multiple subnets, i.e., for any $t \in T_v^u(SN)$: $|\{1 \leq i \leq n \mid t \in T^i\}| \geq 1$, and
- each edge appears in precisely one of the subnets, i.e., for any $(x, y) \in F$: $|\{1 \leq i \leq n \mid (x, y) \in F^i\}| = 1$.

In order to instantiate a decomposition complying with this new definition of valid decomposition, Algorithm 12.1 needs to be refined (cf. Algorithm 13.1).

Algorithm 13.1 Refined k-decomposition algorithm

---

function k-DEC(RPST, k)
    $V = \{root(RPST)\}$
    $D = \emptyset$
    while $V \neq \emptyset$ **do**
        $v \leftarrow pop(V)$
        if $|v.arcs()| \leq k$ **or** not $Decomposable(v)$ **then**
            $D = D \cup \{v\}$
        else $V = V \cup \{children(v)\}$
    return D

function DECOMPOSABLE(s)
    $\{s_1, \ldots s_n\} \leftarrow children(s)$
    $T \leftarrow$ shared transitions in $\{s_1, \ldots s_n\}$
    $P \leftarrow$ shared places in $\{s_1, \ldots s_n\}$
    $T^P \leftarrow$ transitions connected with P

    if $T \cap T_v^u \neq T$ **then return** false
    else if $T^P \cap T_v^u \neq T$ **then return** false
    else if same label in different $\{s_1, \ldots s_n\}$ **then**
        return false
    else return true

Algorithm 13.1 checks if considering the children of a SESE s will violate the definition of valid decomposition in Definition 13.5. The three conditions need to be satisfied:

- transitions shared (T) between any subset of SESEs $\{s_1, \ldots s_n\}$ must be unique visible transitions (T_v^u).

- places shared (P) between any subset of SESEs $\{s_1,\ldots s_n\}$ will be *bridged* according to Def. 13.2. Therefore, transitions connected with the places shared (P) between any subset of $\{s_1,\ldots s_n\}$ must be unique visible transitions (T_v^u), in order to avoid be duplicated boundary transitions after the bridging.
- Transitions with the same label must belong to the same v_i.

The main difference between the original k-decomposition algorithm presented previously and Algorithm 13.1 is that the latter checks if the children of SESE v of the decomposition D are violating the valid decomposition definition (Definition 13.5). Notice that by definition, if the children $\{s_1,\ldots s_n\}$ of v are violating the definition, further descendants of v are considered to be violating the definition as well. Therefore, when the algorithm checks that the SESE must not be decomposed, it includes it into the decomposition D. As a result, Algorithm 13.1 does not guarantee the k property, i.e., some components may have more than k arcs. For instance, consider the subnets resulting from a 15-decomposition and bridging shown in Figure 13.6. Unlike Figure 13.5, here when the algorithm tries to decompose the SESE S_2, it detects that this will result in splitting the duplicate e, and thus it must consider S_2, even though the number of arcs of S_2 is greater 15^1. Notice that some worst case scenarios exist for Algorithm 13.1: consider the example of Figure 13.7. In this case, the presence of invisible transitions in the model boundaries makes it impossible for the algorithm decompose more that the root S_1, and therefore, the resulting decomposition will be the overall net. The effect of those cases can be alleviated by pre-processing the model and the log before applying the decomposed conformance.

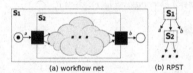

(a) workflow net   (b) RPST

Fig. 13.7: Example of worst case scenario for the k-decomposition with invisible/duplicates.

13.4 Summary

This chapter presented an extension of the SESE decomposition approach to provide a valid decomposition. The valid decomposition is proven to preserve the perfectly fitting condition of the general case, even when the different parts are analyzed inde-

---

[1] Notice that, the bridging may produce that a SESE loses its SESE structure, e.g., the entry and exit places of S_2 are removed when it becomes S_2' due to the bridges B_2 and B_3. In spite of this, the decomposition obtained still satisfies Definition 13.5.

pendently. The following chapters present empirical results and propose tools based on the decomposition topology to help in the diagnosis of conformance problems.

Chapter 14
Decomposing Conformance Checking in Practice

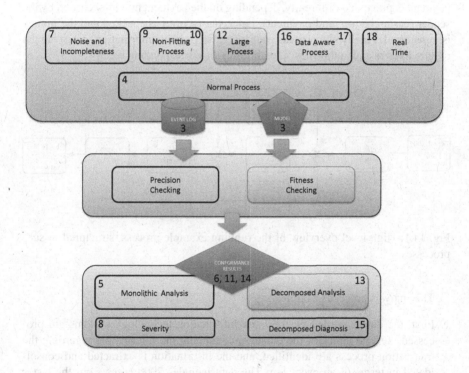

In the previous chapters, we presented a decomposition method to analyze the fitness of large processes. This chapter presents a practical case example where this decomposed conformance checking technique is applied to analyze possible conformance anomalies in a large process and illustrates the detailed diagnostic information that can be obtained. Additionally, the chapter provides an experimental analysis on the different characteristics of the approach.

© Springer International Publishing AG 2016
J. Munoz-Gama: Conf. Check. ... in Process Mining, LNBIP 270, pp. 141–149, 2016.
DOI: 10.1007/978-3-319-49451-7_14

14.1 The Bank Case: The Transaction Process

In this section we propose the analysis of a realistic process case example in order to illustrate the approach proposed in the book. The case involves a bank interested in comparing the real execution of their processes with respect to verified and pre-approved protocols, in order to detect fitness violations of the bank policies.

In particular, this analysis focuses on the *transaction* process of the bank. The transaction process contains all sort of monetary checks, authority notifications, and logging mechanisms responding to the new degree of responsibility and account-ability that current economic environments demand. The process is structured as follows (cf. Figure 14.1 shows a high-level overview of the complete process): it is initiated when a new transaction is requested, opening a new instance in the system and registering all the components involved. The second step is to run a check on the person (or entity) – the origin of the monetary transaction. Then, the actual payment is processed differently, depending on the payment modality chosen by the sender (cash, cheque[1] and payment). Later, the receiver is checked and the money is transferred. Finally, the process ends by registering the information, notifying the required actors and authorities, and emitting the corresponding receipt.

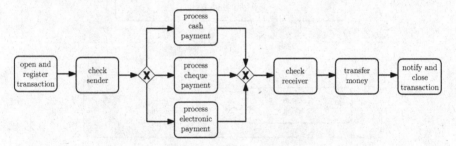

Fig. 14.1: High level overview of the running example process, structured in sub-processes.

The analysis proceeds as follows:

1. First, the data recorded by the bank information technology systems are pro-cessed. Several tables of the database containing the information regarding the transaction process are identified, and the information is extracted and consol-idated in terms of an event log. The data includes 2000 cases from the last 2 months, and 113 different types of events.
2. In the second step, a process model of the transaction process is required. For that, the bank models the process using Petri net notation, as it must be accord-ing to their protocols. Figure 14.2 shows the resulting Petri net, containing 114 transitions.

---

[1] The British term is used to avoid ambiguity with the verb "to check".

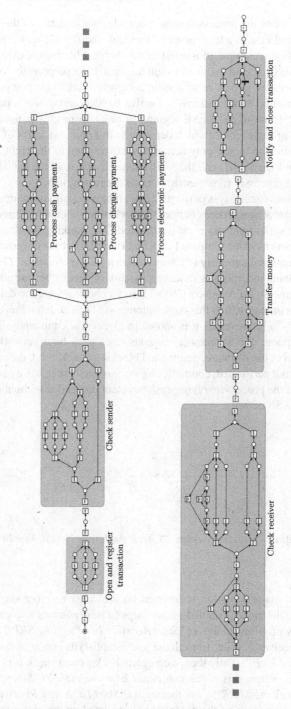

Fig. 14.2: Running example: final valid SESE-decomposition.
The substructures are named according to Figure 14.1.

3. Given the size of the process, and in order to better understand the fitness violations occurred in such a large process, the bank decides to check fitness using the decomposition technique illustrated in this book. The decomposition technique based on SESE regions is used to automatically decompose the overall model into subprocesses. In particular, a valid decomposition where components have a size of at most 60 nodes is derived. Finally, the decomposition is post-processed by merging some of the SESE regions in order to reach the final decomposition shown in Figure 14.2 (which depicts the full process): eight of the proposed subnets (indicated by the gray rectangles) correspond to the eight subprocesses identified in Figure 14.1, and the ninth subnet contains all the trivial connections between subprocesses (represented outside the rectangles).

4. Once the fitness analysis is performed, two major violations are detected:
   First, the transaction process defines that, whenever a client executes a payment in cash, the serial numbers must be checked. The banking regulation states that serial numbers must be compared with an external database governed by a recognized international authority (*"Check Authority Serial Numbers CASN"*). In addition, the bank incorporates two complementary checks: an internal bank check (*"Check Bank Serial Numbers CBSN"*), and a check among the databases of the bank consortium to which this bank belongs to (*"Check Inter-Bank Serial Numbers CIBSN"*). However, as it is shown in Figure 14.3, the analysis reveals that at a given point, due to technical reasons (e.g., peak hour network congestion, malfunction of the software, deliberated blocking attack, etc.), the external check *CASN* was not performed, contradicting the modeled process, i.e., all the running instances of the process involving cash payment proceeded without executing the required CASN activity.

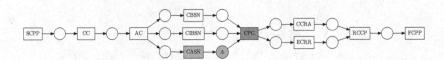

Fig. 14.3: Fitness violation where *"Check Authority Serial Numbers CASN"* is skipped.

The second major violation is detected on the *check receiver* stage of the process, where the model establishes two steps to be performed sequentially: first, a preliminary profiling analysis (*"Start Receiver Pre Profiling SRPP"*) is executed over the receiver in order to evaluate and establish its potential risk (*"Evaluate Pre Profiling EPP"*). Only then, a complete background check is performed over the receiver, where this check can either be more casual (*"Start Low Risk Receiver Processing SLRRP "*) or thorough (*"Start High Risk Receiver Processing SHRRP"*) depending on the potential risk detected on the preliminary profiling. However, the analysis reveals that the evaluation of the receiver is executed with an unfinished preliminary profile check, as it is depicted in in Figure 14.4. This could be produced by the presence of an inexperienced bank employee, malevolence, or simply a badly implemented bank evaluation protocol.

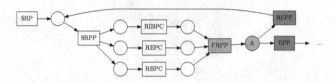

Fig. 14.4: The preliminary profile check for receivers is skipped (*SRPP* to *FRPP*).

5. Once the analysis is concluded, and given that the process model describes the process as it should be executed, the bank decides to analyze in detail the anomalous parts of the process and the people involved, and eventually take the necessary countermeasures on the process execution.

14.2 Experimental Results

In this section we provide experimental results to demonstrate that the proposed decomposition approach provides significant performance gains and improved diagnostics.

Conformance Diagnosis

The goals of decomposed and non-decomposed (i.e., [18]) approaches are slightly different: while [18] aims for a global conformance, the decomposed approach aims for an optimal conformance of each component. However, the decomposed approach proposed in this book makes it possible to identify conformance problems at localized regions. This leads to a better understanding of the causes, aids the diagnosis of conformance problems in large systems, and pinpoints the subprocesses that are producing them. In order to illustrate this contribution, we analyze the fitness results per component for the running example of Chapter 13 and the dataset *bpm2013* using the proposed decomposed approach.

We use a circumference to graphically depict the fitness evaluation of a decomposition by means of a colored gradient for each component. All components of the decomposition are placed in different positions of the circumference.

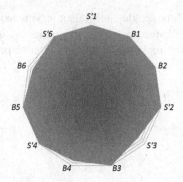

Fig. 14.5: Fitness visualization for the running example.

Let us use the running example of this chapter to illustrate the graphical visualization used.

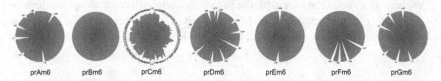

Fig. 14.6: Fitness results per components for benchmark *bpm2013*.

As it is shown in Figure 14.5, for each component, a line from the center of the circumference indicates its fitness. If the line reaches the perimeter, the fitness is 1.0 (components S'_1, B_1, B_2, S'_2, B_3, S'_4, B_5, S'_6), while the line for components with fitness anomalies does not reach the perimeter. To show intuitively the fitness, a color gradient is included in the circumference: the fitness ranges from red/dark (fitness problems close to 0.0) down to green/light (perfect fitness of 1.0).

Figure 14.6 shows the results of the fitness diagnosis of each one of the models of the dataset *bpm2013*: for model *prAm6*, 7 components have fitness anomalies, with diverse severity (7 dents on the circumference) .[2] On the other hand, all components in *prBm6* are perfectly fitting. This contrasts with *prCm6*, where fitness problems are clearly spread over multiple components. The other of dataset model-log pairs have fitness anomalies in just a few components. This supports the approach taken in this book, i.e., the diagnostics help to focus on the problematic parts while at the same time provide performance gains.

---

[2] When no fitness anomalies exist, we do not explicitly label components in the visualization.

Performance Improvements

This experiment is designed to validate the applicability, in terms of performance, of the decomposed alignment-based approach when facing large processes.

For this analysis we use the *bpm2013* dataset[3]. The dataset contains pairs of large models and logs with different levels of fitness (ranging from perfectly fitting as *prBm6*, to pairs with fitness of 0.57 – like *prCm6*), according to the fitness metric in [18]. The analysis includes the computation time of checking fitness with four different approaches, including the Decomposed Conformance (*new DC*) proposed in this book using a k to decompose of 25 and 50 respectively (cf. Algorithm 12.1). For comparison reasons the analysis also includes the results of the old decomposition approach (*old DC*) presented in [64, 63], and the non-decomposed results of [18] technique (*non decomposed*).

Figure 14.7 shows the results of the experiment. The chart illustrates perfectly the vast difference, in computation time, between the presented approach and the non-decomposed alternative. The non-decomposed approach remains competitive for the less complex and highly fitting models (e.g., *prAm6* and *prBm6*). Because of the component creation overhead the non-decomposed approach may even be faster for simple and well-fitting models as noted in [63]. For example, for *prAm6* and *prBm6* the non-decomposed approach is faster than the implementation presented in [63]. This is no longer the case for the new decomposed implementation which is outperforming the earlier approaches. In some cases, the difference could reach two orders of magnitude (e.g., from 15 to 3566 seconds in *prEm6*). More importantly, the proposed approach is able to tackle and provide conformance information for those cases (*prDm6*, *prFm6* and *prGm6*) where [18] is not able to provide a result within a period of 12 hours.

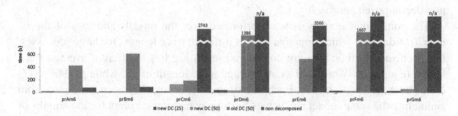

Fig. 14.7: Comparison of computation time among different approaches: the new decomposed conformance checking technique (two variants: one which limits the maximum size of each component to $k = 25$ and the other to $k = 50$), the old decomposed conformance checking technique [63], and the approach without decomposition.

---

[3] http://dx.doi.org/10.4121/uuid:44c32783-15d0-4dbd-af8a-78b97be3de49

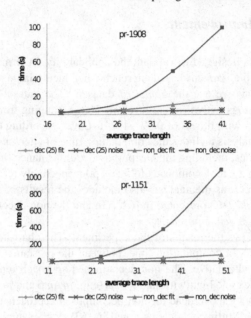

Fig. 14.8: Comparison of computation time among different trace lengths.

Trace length and grouping

This set of experiments was conducted to study the effect that the decomposed approach proposed in the book has on computation time, and to compare it with the non-decomposed approach.

The conformance analysis was performed over the models and logs of the *is-bpm2013* dataset[4] which contains logs of different trace lengths for large processes. For each model, four logs are considered; each log has a different average trace length (e.g., *pr1908-m18-l1* has an average trace length of 18, while *pr1908-m41-l4* has average length of 41). Each one of these four logs has been generated from simulating the same model and using the same parameters (except for the lengths of the traces), and all them are completely fitting. Additionally, another four logs for each model are considered, with the same characteristics, but containing noise (and hence being non-fitting).

Figure 14.8 shows the results for two models included in the dataset: *pr-1908* and *pr-1151*, given that the results are similar for the rest of models-logs in the benchmark. For each model, the chart contains the computation times of each alternative : decomposed using *k* of 25 with noisy logs (*dec(25) noise*) and fitting logs (*dec(25) fit*), and the results for the same noisy (*non_dec noise*) and fitting logs (*non_dec noise*) using the original non-decomposed approach in [18].

---

4 http://dx.doi.org/10.4121/uuid:b8c59ccb-6e14-4fab-976d-dd76707bcb8a

The first conclusion that arises from the experiments refers to the processes with noise – the most plausible assumption in a real world scenario. Figure 14.8 shows that, when the log has short traces, both decomposed and non-decomposed alignment checking perform good. However, once the length of the traces grows, it has a severe effect on the non-decomposed performance. This was to be expected, i.e., the more activities in a trace, the more difficult it is to compute the alignment. On the other hand, the decomposed approach performs both quickly and with a near-to constant growth (and eventually constant at some point). This is caused by the effect of the decomposition on the computation time (as has been shown in Figure 14.7), but also due to the *grouping* (as explained below).

The current implementation of the alignment-based conformance checking includes the grouping optimization: when the algorithm analyzes a trace, it first checks if it has already computed an alignment for an identical trace in the same SESE component. If this is the case, it re-uses the previously computed alignment, thus reducing the time significantly. The effect of this optimization for the non-decomposed scenario depends on the case at hand: the likeliness of identical cases falls as traces get longer, thus a decomposed model has less likely identical traces. The smaller is the component (e.g., $k = 25$), the fewer activities it contains, and therefore, the more likely it is to find a trace already seen before (once the original trace has been projected onto the component). The effects of the grouping are perfectly reflected by the fitting cases (*fit*) of Figure 14.8 where the decomposed approach performs faster than the non-decomposed alternative even in a fitting scenario. This is remarkable because alignments can be created easily in this case.

Chapter 15
Diagnosing Conformance

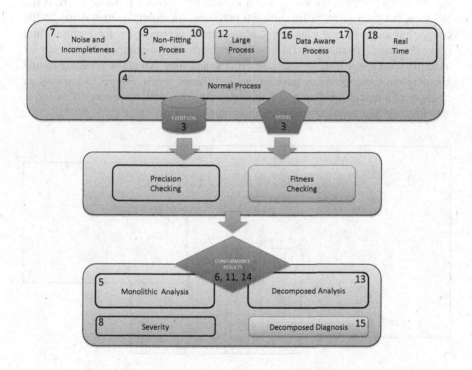

In conformance analysis, computing the conformance metrics is as important as the diagnosis and understanding the cause of the conformance problems. Similar to a map, this chapter provides mechanisms to zoom-in and zoom-out on the conformance analysis, and to detect what parts of the model represent the conformance problems. This technique complements the decomposed approaches presented in previous chapters.

© Springer International Publishing AG 2016
J. Munoz-Gama: Conf. Check. ... in Process Mining, LNBIP 270, pp. 151–161, 2016.
DOI: 10.1007/978-3-319-49451-7_15

15.1 Introduction

The main goal of decomposed conformance checking techniques is to alleviate the time required to analyze conformance, especially for complex alignment-based approaches. This is the case of the approach presented in Chapter 12. However, in conformance analysis, computing the conformance metrics is as important as the diagnosis and understanding the cause of the conformance problems. Approaches that decompose processes into components provide a basic mechanism to diagnose conformance, assessing which components are problematic – especially those techniques that decompose the process into meaningful components such as SESEs [62] or passages [8]. This chapters goes a step further in that direction, and provides additional mechanisms to diagnose conformance problems based on decomposed processes.

Comparing a process model with a map is a recurrent metaphor in process mining [6]. A map represents a city, like a models represents a process. There is not "a" map, but a set of maps for different purposes, for example street maps, traffic congestion maps, or real state maps. Maps can be decomposed, for example, into provinces, cities, or districts. Information in maps can be grouped, to indicate for example what areas concentrate most crimes, and they can allow zooming-in and zooming-out to analyze this information at different levels. Figure 15.1 shows an example of a real estate map which indicates the areas with more rent offers and is able to refine geographically this information with zoom.

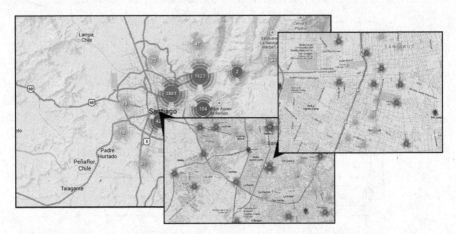

Fig. 15.1: Interactive real estate map from www.portalinmobiliario.com, with grouping by proximity and zoom-in zoom-out functionality.

This chapter translates some of this map functionalities to decomposed conformance checking. In Section 15.2 we define a topological relationship between components, and use this relation to define close areas with conformance problems. Section 15.3 defines a hierarchy among components, opening the door to explore the conformance results at multiple levels. Moreover, the section proposes several refinements over the hierarchy that helps the conformance diagnosis. Finally, Section 15.4 provides some experimental results, and Section 15.5 concludes the chapter.

15.2 Topological Conformance Diagnosis

A valid decomposition, presented in the previous chapter, is a collection of subnets that may be related to each other through the sharing of transitions, i.e., two subnets are related if they share a transition. The *topology* of a valid decomposition is an undirected graph where the vertices denote subnets and the edges denote the sharing of transitions.

Definition 15.1 (Topology of a Decomposition) *Let $D = \{SN^1, SN^2, \ldots SN^n\}$ be a valid decomposition, where $SN^i = (PN^i, M_I^i, M_F^i)$ and $PN^i = (P^i, T^i, F^i)$. The topology of decomposition D is defined as the undirected graph $T_D = (D, C)$ such that two components are connected if they share any transition, i.e., $C = \{\{SN^i, SN^j\} | 1 \leq i < j \leq n \wedge T^i \cap T^j \neq \emptyset\}$.*

In the general definition of topology over a valid decomposition the relations remain undirected, i.e., two subnets sharing the same transition are connected by an undirected edge. However, in the specific case of a valid decomposition derived from SESEs, defined in Chapter 12, this definition can be extended to include the concept of direction: the transition being the exit of the SESE is considered the source of the edge, while the entry is the target. Bridges can have multiple entry and exit nodes, but again we can derive the direction connections among bridges and SESEs.

Definition 15.2 (Topology of a SESE Decomposition) *Let $\mathbb{D} = \{S_1, \ldots S_n\}$ and $\mathbb{D}' = \{S_1', \ldots S_n', B_1, \ldots B_k\}$ be a SESE decomposition before and after applying bridging. Let $\{p_1, \ldots, p_k\}$ be the boundary places in $\mathbb{D}$. Let $D_{\mathbb{D}'} = \{SN^{S_1'}, \ldots SN^{S_n'}, SN^{B_1} \ldots SN^{B_k}\}$ represent the decomposition constructed from $\mathbb{D}'$. The topology of $D_{\mathbb{D}'}$ is defined as the directed graph $T_{D_{\mathbb{D}'}} = (D_{\mathbb{D}'}, C)$ such that $C = \{(SN^{S_i'}, SN^{S_j'}) | 1 \leq i, j \leq n \wedge (y, x) \in S_i \wedge (x, z) \in S_j\} \cup \{(SN^{S_i'}, SN^{B_j}) | 1 \leq i \leq n \wedge 1 \leq j \leq k \wedge (y, p_j) \in S_i\} \cup \{(SN^{B_j}, SN^{S_i'}) | 1 \leq i \leq n \wedge 1 \leq j \leq k \wedge (p_j, y) \in S_i\}$.*

Note that the topological graph has as many vertices as the nets in $\mathbb{D}'$, but some of the arcs in this graph (those regarding connection to bridges) are defined over the original SESE decomposition $\mathbb{D}$, e.g., $(y, p_j) \in S_i$ refers to an arc in the original SESE and is used to infer a directed connection from SN^{S_i} to SN^{B_j}.

One of the features of the topology is to aid the visualization of a valid decomposition. For example, let us consider the valid decomposition in Figure 15.2 (a slight modification of the model in Figure 12.1 in Chapter 12). The decomposition is the result of applying a 4-decomposition over the model (i.e., SESEs with at most 4

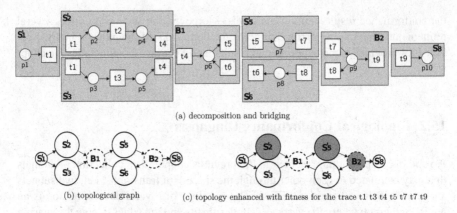

(a) decomposition and bridging

(b) topological graph   (c) topology enhanced with fitness for the trace t1 t3 t4 t5 t7 t7 t9

Fig. 15.2: Example of valid decomposition and its topology

edges: $S_1', S_2', S_3', S_5', S_6', S_8'$) and followed by the bridging (resulting in two bridges, B_1 and B_2, corresponding with the two boundary places p_6 and p_9)[1]. The corresponding topology is shown in same Figure 15.2b.

Besides simply showing the connections among subnets, the topology can be enhanced with other information about the components and their characteristics. For instance, bridges can be denoted by circles having dotted borders and SESEs can be denoted by circles having solid borders. Moreover, the size of the nodes in the graph is directly related with the size of the corresponding subnets, i.e., a subnet with many arcs is depicted using a larger circle compared to subnets with fewer arcs. Given the final goal of this thesis (i.e., conformance analysis), a particular interesting case is to enhance the topology with conformance information. For example, let us consider the trace $\sigma = t_1 t_3 t_4 t_5 t_7 t_7 t_9$. When we check fitness in the subnets of decomposition $D_{\mathbb{D}'} = \{SN^{S_1'}, \dots SN^{S_8'}, SN^{B_1}, SN^{B_2}\}$, we detect the following fitness anomalies: in $SN^{S_2'}$, t_4 is fired without firing t_2; in $SN^{S_5'}$, t_7 is executed twice, but this requires t_5 to be fired twice as well; finally, in the bridge SN^{B_2}, t_7 is fired twice, but t_9 only once, leaving a token remaining in p_9. This information can be used to enhance the topology. As shown in Figure 15.2c the vertices which have problems can be depicted in color (here S_2', S_5' and B_2).

Although the topology is an important aid for the process diagnosis by itself, it can also guide further analysis. For instance, the topological graph enhanced with conformance information can be used to *identify maximal process fragments with fitness problems*. This allows us to focus on the problematic parts of a model, discarding the parts without any fitness problems. Algorithm 15.1 describes a procedure that is based on detecting connected components (C_c) on the graph induced by the non-fitting vertices. First, the topological graph (T_D) is filtered, leaving only non-fitting vertices (V). Then, the weakly connected components (C_c) are detected:

---

[1] Note that the original trivial SESE S_4 that corresponds to the arc $(t4, p6)$ has disappeared once the bridging has been done, i.e., the arc is now in B_1. The same happens for the original trivial SESE S_7 corresponding to the arc $(p9, t9)$.

1) a random node v_1 is chosen, 2) all nodes $\{v_1, \ldots v_n\}$ weakly connected (i.e., connected vertices without considering the direction of the edges) with v_1 are computed using a depth-first search exploration and they constitute a new connected component, and finally 4) $\{v_1, \ldots v_n\}$ are removed from the graph and the exploration of connected components continues. For each connected component, we project the elements of the original net that they refer to. Note that this algorithm prioritizes the connectivity among vertices resulting in weakly connected components. However, alternative versions of the algorithm that yield strongly connected components are possible. For instance, given the example of Figure 15.2c, two connected components are found as shown in Figure 15.3: one corresponding to $SN^{S_2'}$ and the other to the union of $SN^{S_5'}$ and SN^{B_2}.

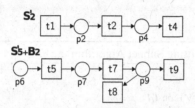

Fig. 15.3: Examples of non-fitting weakly connected components.

Algorithm 15.1 Non-Fitting Weakly Connected Components Algorithm

---

function NFWCC(T_D, V)   ▷ V is non-fitting vertices
    $C_c = \emptyset$
    remove from T_D: ▷ Graph induced by V
    -all arcs $c = \{x, y\}$ such that $x, y \notin V$
    -all vertices $z \notin V$

    while T_D has vertices **do**
        $v_1 \leftarrow$ select random vertex on T_D
        $\{v_1, \ldots v_n\} \leftarrow$ get vertices weakly connected with v_1 using Depth-first search
        remove $\{v_1, \ldots v_n\}$ from T_D
        $C_c = C_c \cup \bigcup_1^n v_i$
    return C_c

The topological graph enhanced with conformance information can also be used to create one complete subnet that includes all non-fitting subnets of the decomposition, i.e., a *connected* set of vertices V containing all the non-fitting vertices V_{nf}. Algorithm 15.2 illustrates the heuristic-based approach proposed, based on the greedy expansion of the largest non-fitting connected components, to compute the complete non-fitting subnet. Initially, V contains the non-fitting vertices V_{nf}, and G denotes the graph induced by V. The goal of the algorithm is to have all the vertices in V connected, i.e. G be connected. If this is not the case, the algorithm detects the two largest connected components (c_1 and c_2) of G, and then computes the shortest path ($\{v_1 \ldots v_n\}$) between any vertex in c_1 and any vertex in c_2. Finally, $\{v_1 \ldots v_n\}$ are included to V, and it is checked again if the induced graph G is connected. Given the example of Figure 15.2c, the net resulting (shown in Figure 15.4) contains the union of the subnets $SN^{S'_2}$, SN^{S_4}, SN^{B_1}, $SN^{S'_5}$ and SN^{B_2}.

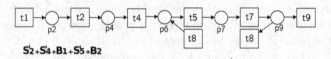

S'2+S'4+B1+S'5+B2

Fig. 15.4: Example of a non-fitting subnet.

Algorithm 15.2 Non-Fitting Subnet Algorithm

function NFN(T_D, V_{nf})   ▷ V_{nf} is non-fitting vertices
   $V \leftarrow V_{nf}$
   $G \leftarrow$ graph induced by V on T_D
   while G is not connected **do**
      $c_1 \leftarrow$ get the $1st$ largest conn. comp. of G
      $c_2 \leftarrow$ get the $2nd$ largest conn. comp. of G
      $\{v_1 \ldots v_n\} \leftarrow$ shortest_path_vertex(T_D, c_1, c_2)
      $V = V \cup \{v_1 \ldots v_n\}$.
      $G \leftarrow$ graph induced by V on T_D
   return Petri net induced by V

15.3 Multi-level Conformance Diagnosis and its Applications

So far the analysis of the conformance was always performed using a complete decomposition of the model. However, for detailed process diagnosis it is important to also be able to do a more focused analysis. This section presents three approaches to achieve this: (1) stand-alone checking, (2) multi-level analysis, and (3) filtering.

15.3.1 Stand-alone Checking

First we consider the problem of analyzing a selected subprocess in isolation. Clearly, assumptions on the subprocess and its context must be defined in order to perform such an isolated conformance check. The conformance results obtained are strongly correlated with the assumptions considered, and hence the analysis of the model properties and domain knowledge becomes an essential part, e.g., whether a place has a bound on the number of tokens, or the number of activations of the subprocess within a trace.

Let us show an application of the stand-alone checking for the typical case of well-structured process models, that can easily be modeled using the subclass of *safe workflow nets*[64]. Given a SESE S obtained from a decomposition, one can apply the following steps to conduct a local diagnosis of S:

1. *Workflowing the SESE:* In order to have a clear starting and ending point for the subprocess represented, re-define the net derived from S. In other words, given a SESE S, define the net derived from S in terms of a workflow net, with an starting place (*start*) and a final place (*end*). By construction, a SESE has both an entry (i) and an exit (o) node. The *start* corresponds with i if i is a place. However, when i is a transition, we define *start* to be an artificial place and we connect it with i. Similarly for *end* and o.

2. *Initial and Final Marking:* Given the workflow-net from the previous step, determining a plausible initial marking becomes straightforward, i.e., due to the safeness assumption of safe workflow nets, we consider a single token in the *start* in order to *enable* the execution of the subprocess. Similarly for the final marking.

3. *SESE activations:* the number of potential activations of a SESE within a case must be determined. In the case where it is always one, the SESE is left as is. However, in the case where it can be executed more than once (e.g., the SESE is inside some loop in the model), the net in the previous step is *short-circuited*, using a silent transition between *end* and *start*. Finally, it can also happen that a SESE may be not executed in a trace. In this last case, a silent transition between *start* and *end* avoiding the SESE content will be used. Determining if a suprocess can be executed several times is a complex matter. In [64], it is proposed the use of Petri net structural theory (*minimal T-invariants* [82]) as a best effort strategy for estimating repetitive behavior.

15.3.2 Multi-Level Analysis

In this section we propose to combine the stand-alone checking presented above with the RPST to achieve conformance analysis in a hierarchical manner. RPST nodes enriched with conformance information enable analysis at different degrees of granularity and independence, similar to zooming in and out using online maps. Note that, by construction, the root of the RPST is the overall net. Therefore, any hierarchical analysis that involves the conformance checking of all the RPST nodes will require checking conformance on the original net (plus the checks of the rest of nodes), i.e., the computation time for a exhaustive hierarchical analysis will always be, by definition, greater than checking conformance on the overall net. For complex and time-consuming cases, this problem can be alleviated by limiting the size of the nodes to check or by using less expensive replay-based conformance techniques like [77, 28]. The latter techniques use heuristics in order to deal with unfitting situations.

15.3.3 Filtering

The study presented in [64] suggest that there are three main differences between manual hierarchical decomposition and the one provided by the RPST-based decomposition: (1) analysts prefer to *discard small components*, (2) analysts prefer to *not consider similar components*, and (3) analysts prefer to *have a hierarchy with a limited number of levels*. Additionally, in this paper we point out a fourth difference: (4) analysts prefer to *base hierarchies on other (non-control-flow) perspectives*. In the remainder of this section we propose filtering techniques to allow for RPST-based decompositions closer to hierarchical decompositions preferred by analysts.

- *Small components:* Small components of the RPST can be removed by filtered using a minimal size threshold.
- *Similarity:* In order to reduce the redundancy of components and the unnecessary growth of the hierarchy, a *similarity metric* between parent-child components is defined, together with a threshold that determines the similarity frontier that will determine when two components are considered essentially the same. The proposed metric for estimating the similarity between a node S and its *single* child S' is based on two factors: size and simplicity. The *size* factor is related with the number of arcs of S not included on S'. The more arcs shared by both components, the more similar they are. For instance, considering the component S_1 of Figure 15.5a, all its arcs are included in S_2 except two, i.e., S_2 is in essence S_1. Therefore, a detailed conformance diagnosis over S_1 may be sufficient for understanding both subprocesses. The *simplicity* factor refers to the simplicity of part of the parent S not included on the child S'. When such part defines a simple behavior (e.g., the strictly sequential behavior of S_3 not included in S_4, in Figure 15.5b), the analysis and understanding of the parent may again be enough. On the other hand, when the behavior not included in S' contains complex control-flow constructs (e.g., mixtures of concurrency and choice) it may be more advisable to analyze both subprocesses. An example similarity metric is formalized as follows.

Definition 15.3 (Similarity Metric) *Let $S_P = (V_P, F_P)$ be an RPST node, and let $S_C = (V_C, F_C)$ be its only child. Let* size *define the difference on size between them, i.e., $size = |F_C|/|F_P|$. Let $F_O = F_P \setminus F_C$ be the set of non-intersecting arcs. Let F_O^* be the arcs in F_O that have a source vertex with only one outgoing edge, and a target vertex with only one incoming edge, i.e., $F_O^* = \{(x,y) \in F_O | (x,v) \in F_O| = 1 \wedge |(w,y) \in F_O| = 1\}$. Let* simplicity *define the simplicity of the non-intersecting arcs, i.e., $simplicity = |F_O^*|/|F_O|$. The similarity between S_P and S_C is the harmonic mean between size and simplicity:*

$$similarity = 2 \cdot \frac{size \cdot simplicity}{size + simplicity}$$

Although the similarity evaluation is restricted to nodes with only one child, our experimental results show that the reduction achieved on the RPST may be significant (especially after applying a small nodes filtering).

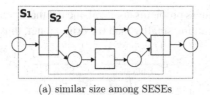

(a) similar size among SESEs

(b) high simplicity among SESEs

Fig. 15.5: Example of cases with high similarity between nested SESEs.

- *Multi-perspective filtering:* The filtering presented until now is based on only structural net properties and does not take into account other perspectives (e.g., data, costs, roles, departments). However, there may be situations where we would like to focus the analysis only on subprocesses which satisfy certain domain conditions, e.g., an analyst may want to focus on the subprocesses that involve tasks executed in a particular department. Therefore, we need to support filtering based on user-requirements and focus the analysis on the subprocesses involving activities relevant from the selected viewpoint. Such filtering is not limited to activities and may involve other perspectives (e.g., resources, actors, or costs), determining the activities they are connected with, and using them for filtering.

15.4 Experimental Results

The set of experiments of this section is designed to illustrate the effects of some of the techniques proposed for process diagnosis. In particular, the *Non-fitting Subnet Algorithm* (cf. Algorithm 15.2), and the techniques of filtering the RPST based on small components and similarity (cf. Section 15.3.3). Table 15.1 shows the application of the NFN algorithm over the benchmark *bpm2013*[2], with components of size at most 50. For each model (containing P places and T transitions) the table provides the size of the minimal net containing all the non-fitting components, i.e., the number of places and transitions ($|P|$ and $|T|$), and the number of vertices $|V|$ used to create the net. The table illustrates the benefits of the proposed algorithm to detect and isolate the fitness mismatches. In cases where the fitness problems are spread all over the whole model, the resulting net is almost the original net (e.g., *prCm6*).

---

[2] http://dx.doi.org/10.4121/uuid:44c32783-15d0-4dbd-af8a-78b97be3de49

However, when the fitness problems are local, the net that encloses all problem spots may be orders of magnitude smaller than the original net, thus easing the diagnosis.

Dataset		NFN		
P	T	$\|V\|$	$\|P\|$	$\|T\|$
prAm6 363	347	14	15	14
prCm6 317	317	113	315	317
prDm6 529	429	31	55	52
prEm6 277	275	31	29	40
prFm6 362	299	7	27	25
prGm6 357	335	5	34	29

Table 15.1: Results of NFN algorithm.

The second experiment performed illustrates the effects of the simplification techniques over the RPST. Figure 15.6 reflects the results for one of the models (*prCm*6). The charts show the number of nodes of the original RPST, after filtering small components (< 10) and then merging by similarity (> 0.8). The number of nodes are distributed by levels of depth in the RPST tree, i.e., the distance with the root represented as the level 1. The chart clearly reflects the difference between the number of components on the original RPST and the one after removing the small components, i.e., most of the RPST nodes are small. After removing small nodes the depth of the RPST only decreases two levels (from 14 to 12). On the other hand, when merging on similarity is applied over the filtered RPST, though the number of nodes is not reduced drastically, there is a significant reduction in the number of levels in the tree(from 13 to 6). This provides a hierarchical decomposition with less redundancy and it is more aligned with the human perception [64].

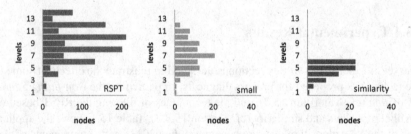

Fig. 15.6: Results of filtering by small (< 10) and merging by similarity (> 0.8) over the model *prCm*6.

15.5 Summary

Decomposition techniques in conformance checking provide an efficient mechanism to identify and locate conformance anomalies. This chapter went a step further, using decomposition techniques to provide other mechanisms for the diagnosis of conformance problems. The chapter proposed a topological relation between components, used to identify closely connected components with conformance problems. Furthermore, the chapter defined a hierarchy of components, opening the door to a zoom-in zoom-out analysis of the conformance.

Chapter 16
Data-aware Processes and Alignments

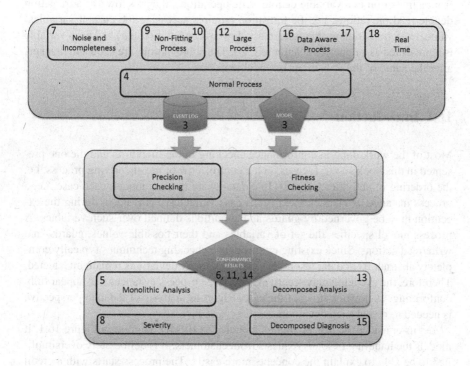

Previous chapters focused on the control-flow perspective of the processes. This chapter explores the extension of the concepts of conformance and alignments beyond the control-flow perspective. In particular, the chapter presents conformance checking for data-aware processes, i.e., process with both control-flow and data perspectives. In the next chapter, we will adapt this approach to handle large processes.

© Springer International Publishing AG 2016
J. Munoz-Gama: Conf. Check. ... in Process Mining, LNBIP 270, pp. 163–172, 2016.
DOI: 10.1007/978-3-319-49451-7_16

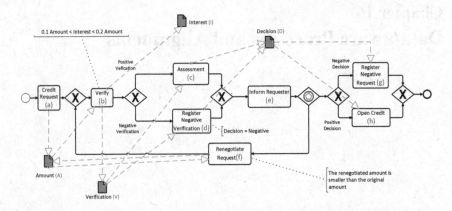

Fig. 16.1: Example of a (simplified) process to request loans. The dotted arcs going from a transition to a variable denote write operations; the arcs towards a transition denote read operations, i.e. the transition requires accessing the current variables' value. Each transition is abbreviated as a lower-case letter (e.g. *a*) and each variable is represented as a upper-case letter (e.g. *A*). The abbreviations are shown in brackets after the name of the transitions or variable names.

16.1 Introduction

Most of the work done in conformance checking in the literature, and the one presented in this book so far, focuses on the control-flow of the underlying process, i.e. the ordering of activities [77, 18]. In a data-aware process model, each case, i.e. a process instance, is characterized by its case variables. Paths taken during the execution may be governed by guards and conditions defined over such variables. A process model specifies the set of variables and their possible values, guards, and write/read actions. Since existing conformance checking techniques typically completely abstract from data, resources, and time, many deviations remain undetected. Therefore, the event log may record executions of process instances that appear fully conforming, even when it is not the case. Rigorous analysis of the data perspective is needed to reveal such deviations.

Let us consider the process that is modeled as BPMN diagram in Figure 16.1. It models the handling of loans requests from customers. It is deliberately oversimplified to be able to explain the concepts more easily. The process starts with a credit request where the requestor provides some documents to demonstrate the capability of paying back the loan. These documents are verified and the interest amount is also computed. If the verification step is negative, a negative decision is made, the requestor is informed and, finally, the negative outcome of the request is stored in the system. If verification is positive, an assessment is made to take a final decision. Independent of the assessment's decision, the requestor is informed. Moreover, even if the verification is negative, the requestor can renegotiate the loan (e.g. to have lower interests) by providing further documents or by asking for a smaller amount. In this case, the verification-assessment part is repeated. If both the decision and verification are positive and the requestor is not willing to renegotiate, the credit is opened. Let us consider the following trace:[1]

$$\sigma_{ex} = \langle(\mathbf{a}, \emptyset, \{(A, 4000)\}), (\mathbf{b}, \{(A, 4000)\}, \{(I, 450), (V, \mathsf{false})\}), (\mathbf{c}, \{(V, \mathsf{false})\}, \\ \{(D, \mathsf{true})\}), (\mathbf{e}, \emptyset, \emptyset), (\mathbf{f}, \{(A, 4000)\}, \{(A, 5000)\}), (\mathbf{b}, \{(A, 5000)\}, \{(I, 450), \\ (V, \mathsf{false})\}), (\mathbf{d}, \{(V, \mathsf{false})\}, \{(D, \mathsf{false})\}), (\mathbf{e}, \emptyset, \emptyset), (\mathbf{h}, \{(D, \mathsf{true})\}, \emptyset)\rangle$$

Seen from a control-flow perspective only (i.e. only considering the activities' ordering), the trace seems to be fully conforming. Nonetheless, a number of deviations can be noticed if the data perspective is considered. First of all, if activity c is executed, previously activity b could not have resulted in a negative verification, i.e. V is set to false. Second, activity f cannot write value 5000 to variable A, as this new value is larger than the previous value, i.e. 4000. Furthermore, if the decision and verification are both negative, i.e. both V are D are set to false, then h cannot be executed at the end.

The identification of non-conforming traces clearly has value in itself. Nonetheless, organizations are often interested in explanations that can steer measures to improve the quality of the process. *Alignments* (cf. Chapter 9) aim to support more refined conformance checking. An alignment aligns a case in the event log with an execution path of the process model as good as possible. If the case deviates from the model, then it is not possible to perfectly align with the model and a best matching scenario is selected. Note that for the same deviation, multiple explanations can be given. For instance, the problem that h was executed when it was not supposed to happen can be explained in two ways: (1) h should not have occurred because V and D are both set to false ("control-flow is wrong") and (2) V and D should both have been set to true because h occurs ("data-flow is wrong"). In order to decide for the most reasonable explanation, costs are assigned to deviations and we aim to find the explanation with the lowest cost. For instance, if assigning a wrong value to V and D is less severe than executing h wrongly, the second explanation is preferred. The seminal work in [18] only considers alignments in the control-flow part, thus ignores the data-perspective aspect of conformance.

As we detail in Section 16.2.3, finding an alignment of an event log and a data-aware process model is undecidable in the general case. However, to make the problem decidable, works [50, 53] put forward the limitation that guards need to be linear (in)equations. These works also show that, even with that limitation, the problem of finding an alignment of an event log can become intractable since the problem's complexity is exponential on the size of the model, i.e. the number of activities and data variables.

---

[1] Notation $(\mathbf{act}, r, w)$ is used to denote the occurrence of activity act that writes and reads variables according to functions w and r, e.g., $(\mathbf{b}, \{(A, 4000)\}, \{(I, 450), (V, \mathsf{false})\})$ is an event corresponding to the occurrence of activity $\mathbf{b}$ while reading value 4000 for variable A and writing values 450 and false to variables I and V respectively. $(\mathbf{e}, \emptyset, \emptyset)$ corresponds to the occurrence of activity $\mathbf{e}$ without reading/writing any variables.

16.2 Data-aware Processes

In the previous chapters, all the definitions and theory presented focus on the control-flow perspective of the processes. De Leoni, Mannahardt, Reijers, and van der Aalst extended the existing process mining theory to incorporate the data-perspective [50, 51, 53] In this section we present those concepts. In particular, this section presents Petri nets with data as data-aware process model notation, event logs with data, and the relation between models and logs.

16.2.1 Petri nets with Data

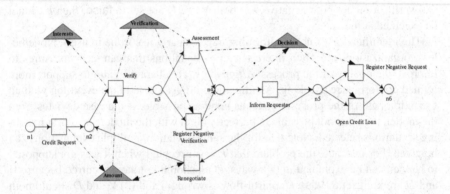

Fig. 16.2: Pictorial representation of a Petri net with Data that models the process earlier described in terms of BPMN diagram (cf. Figure 16.1). Places, transitions and variables are represented as circles, rectangles and triangles, respectively. The dotted arcs going from a transition to a variable denote the writing operations; the reverse arcs denote the read operations, i.e. the transition requires accessing the current variables' value.

Petri nets presented in previous chapters are extended to incorporate data. A Petri net with Data is a Petri net with any number of variables (see Definitions 16.1 and 16.2 below). Petri nets with data can be seen as an abstracted version of high-level/colored Petri nets [48]. Colored Petri nets are extremely rich in expressiveness; however, many aspects are unimportant in our setting. Petri nets with data provide precisely the information needed for conformance checking of data-aware models and logs. In particular, the definitions presented is based on the work of de Leoni et al. [50].

Definition 16.1 (Variables and Values) $\mathcal{U}_{VN}$ *is the universe of variable names.* $\mathcal{U}_{VV}$ *is the universe of values.* $\mathcal{U}_{VM} = \mathcal{U}_{VN} \nrightarrow \mathcal{U}_{VV}$ *is the universe of variable mappings.*

In this type of nets, transitions may read from and/or write to variables. Moreover, transitions are associated with guards over these variables, which define when these they can fire. A guard can be any formula over the process variables using relational operators $(<, >, =)$ as well as logical operators such as conjunction ($\wedge$), disjunction ($\vee$), and negation ($\neg$). A variable v appear as v_r or v_w, denoting the values read and written by the transition for v. We denote with $Formulas(V)$ the universe of such formulas defined over a set V of variables. In the remainder, given a set $V \subset \mathcal{U}_{VN}$ of variable names, we denote $V_R = \{v_r : v \in V\}$ and $V_W = \{v_w : v \in V\}$.

Formally, a *Petri net with Data* (DPN) is defined as follows:

Definition 16.2 (Petri net with Data) *A Petri net with Data DPN $= (SN, V, val, init, read, write, guard)$ consists of*

- *a system net $SN = (PN, M_{init}, M_{final})$ with $PN = (P, T, F, l)$,*
- *a set $V \subseteq \mathcal{U}_{VN}$ of data variables,*
- *a function $val \in V \rightarrow \mathscr{P}(\mathcal{U}_{VV})$ that defines the values admissible for each variable, i.e., $val(v)$ is the set of values that variable v can have,*
- *a function $init \in V \rightarrow \mathcal{U}_{VV}$ that defines the initial value for each variable v such that $init(v) \in val(v)$ (initial values are admissible),*
- *a read function $read \in T \rightarrow \mathscr{P}(V)$ that labels each transition with the set of variables that it reads,*
- *a write function $write \in T \rightarrow \mathscr{P}(V)$ that labels each transition with the set of variables that it writes,*
- *a guard function $guard \in T \rightarrow Formulas(V_W \cup V_R)$ that associates a guard with each transition such that, for any $t \in T$ and for any $v \in V$, if v_r appears in $guard(t)$ then $v \in read(t)$ and if v_w appears in $guard(t)$ then $v \in write(t)$.*

$\mathcal{U}_{DPN}$ *is the* universe of Petri nets with data.

The notion of bindings is essential for the remainder. A *binding* is a triplet (t, r, w) describing the execution of transition t while reading values r and writing values w. A binding is valid if:

1. $r \in read(t) \rightarrow \mathcal{U}_{VV}$ and $w \in write(t) \rightarrow \mathcal{U}_{VV}$
2. for any $v \in read(t)$: $r(v) \in val(v)$, i.e., all values read should be admissible,
3. for any $v \in write(t)$: $w(v) \in val(v)$, i.e., all values written should be admissible.
4. Guard $guard(t)$ evaluates true.

More specifically, let us introduce variable assignment $\chi_b : (V_R \cup V_W) \nrightarrow \mathcal{U}_{VV})$ which is defined as follows: for any $v \in read(t)$, $\chi(v_r) = r(v)$ and, for any $v \in write(t)$, $\chi(v_w) = w(v)$. A binding (t, r, w) makes $guard(t)$ evaluate true if the evaluation of $guard(t)$ wrt. χ_b returns true.

A marking (M, s) of a Petri net with Data *DPN* has two components: $M \in \mathcal{B}(P)$ is the *control-flow marking* and $s \in \mathcal{U}_{VM}$ with $dom(s) = V$ and $s(v) \in val(v)$ for all $v \in V$ is the *data marking*. The initial marking of a Petri net with Data *DPN*

Transition	Guard
Credit Request	true
Verify	$0.1 \cdot A_r < I_w < 0.2 \cdot A_r$
Assessment	$V_R = \text{true}$
Register Negative Verification	$V_r = \text{false} \wedge D_w = \text{false}$
Inform Requester	true
Renegotiate Request	$V_r = \text{false} \wedge A_w < A_r$
Register Negative Request	$D_r = \text{false}$
Open Credit	$D_r = \text{true}$

Table 16.1: Definitions of the guards of the transitions in Fig. 16.2. Variables and transition names are abbreviated as described in Figure 16.1. Subscripts r and w refer to, respectively, the values read and written for that given variable.

is $(M_{init}, init)$. Recall that $init$ is a function that defines the initial value for each variable.

$(DPN, (M, s))[b\rangle$ denotes that a binding b is enabled in marking (M, s), which indicates that each of its input places $\bullet t$ contains at least one token (control-flow enabled), b is valid and and $s\!\upharpoonright_{read(t)} = r$ (the actual values read match the binding).

An enabled binding $b = (t, r, w)$ may *occur*, i.e., one token is removed from each of the input places $\bullet t$ and one token is produced for each of the output places $t\bullet$. Moreover, the variables are updated as specified by w. Formally: $M' = (M \setminus \bullet t) \uplus t\bullet$ is the control-flow marking resulting from firing enabled transition t in marking M (abstracting from data) and $s' = s \oplus w$ is the data marking where $s'(v) = w(v)$ for all $v \in write(t)$ and $s'(v) = s(v)$ for all $v \in V \setminus write(t)$. $(DPN, (M, s))[b\rangle(DPN, (M', s'))$ denotes that b is enabled in (M, s) and the occurrence of b results in marking (M', s').

Figure 16.2 shows a Petri net with Data DPN_{ex} that models the same process as represented in Figure 16.1 as BPMN diagram, and Table 16.1 illustrates the conditions of the guards of the transitions of DPN_{ex}. The labeling function l is such that the domain of l is the set of transitions of DPN_{ex} and, for each transition t of DPN_{ex}, $l(t) = t$. In other words, the set of activity labels coincides with the set of transitions.

Let $\sigma_b = \langle b_1, b_2, \ldots, b_n \rangle$ be a sequence of bindings. $(DPN, (M, s))[\sigma_b\rangle(DPN, (M', s'))$ denotes that there is a set of markings $(M_0, s_0), (M_1, s_1), \ldots, (M_n, s_n)$ such that $(M_0, s_0) = (M, s)$, $(M_n, s_n) = (M', s')$, and $(DPN, (M_i, s_i))[b_{i+1}\rangle(DPN, (M_{i+1}, s_{i+1}))$ for $0 \leq i < n$. A marking (M', s') is *reachable* from (M, s) if there exists a σ_b such that $(DPN, (M, s))[\sigma_b\rangle(DPN, (M', s'))$.

$\phi_f(DPN) = \{\sigma_b \mid \exists_s (DPN, (M_{init}, init))[\sigma_b\rangle(DPN, (M_{final}, s))\}$ is the *set of complete binding sequences*, thus describing the behavior of DPN.

Given a set of Petri nets with Data, the union is defined as the *merge* of those Petri nets.

Definition 16.3 (Union of Petri nets with Data) *Let* $DPN^1 = (SN^1, V^1, val^1, init^1, read^1, write^1, guard^1)$ *and* $DPN^2 = (SN^2, V^2, val^2, init^2, read^2, write^2, guard^2)$ *with* $V^1 \cap V^2 = \emptyset$. $DPN^1 \cup DPN^2 = (SN^1 \cup SN^2, V^1 \cup V^2, val^1 \oplus val^2, init^1 \oplus init^2, read^3, write^3, guard^3)$ *is the union such that*

- $read^3(t) = read^1(t)$, $write^3(t) = write^1(t)$, and $guard^3(t) = guard^1(t)$ if $t \in T^1 \setminus T^2$,
- $read^3(t) = read^2(t)$, $write^3(t) = write^2(t)$, and $guard^3(t) = guard^2(t)$ if $t \in T^2 \setminus T^1$, and
- $read^3(t) = read^1(t) \cup read^2(t)$, $write^3(t) = write^1(t) \cup write^2(t)$, and $guard^3(t) = guard^1(t) \cdot guard^2(t)$ if $t \in T^1 \cap T^2$.

16.2.2 Event Logs and Relating Models to Event Logs

Next we extend the definition of *event logs* presented in Chapter 3 to incorporate data and relate them to the *observable* behavior of a *DPN*.

Definition 16.4 (Trace, Event Log with Data) *A trace* $\sigma \in (\mathcal{U}_A \times \mathcal{U}_{VM} \times \mathcal{U}_{VM})^*$ *is a sequence of activities with input and output data.* $L \in \mathcal{B}((\mathcal{U}_A \times \mathcal{U}_{VM} \times \mathcal{U}_{VM})^*)$ *is an event log with read and write information, i.e., a multiset of traces with data.*

Definition 16.5 (From Bindings to Traces) *Consider a Petri net with Data with transitions* T *and labeling function* $l \in T \nrightarrow \mathcal{U}_A$. *A binding sequence* $\sigma_b \in (T \times \mathcal{U}_{VM} \times \mathcal{U}_{VM})^*$ *can be converted into a trace* $\sigma_v \in (\mathcal{U}_A \times \mathcal{U}_{VM} \times \mathcal{U}_{VM})^*$ *by removing the bindings that correspond to unlabeled transitions and by mapping the labeled transitions onto their corresponding label.* $l(\sigma_b)$ *denotes the corresponding trace* σ_v.

Note that we overload the labeling function to binding sequences, $\sigma_v = l(\sigma_b)$. This is used to define $\phi(DPN)$: the set of all visible traces.

Definition 16.6 (Observable Behavior of a Petri net with Data) *Let DPN be a Petri net with Data.* $(DPN, (M, s))[\sigma_v \triangleright (DPN, (M', s'))$ *if and only if there is a sequence* σ_b *such that* $(DPN, (M, s))[\sigma_b\rangle(DPN, (M', s'))$ *and* $\sigma_v = l(\sigma_b)$. $\phi(DPN) = \{l(\sigma_b) \mid \sigma_b \in \phi_f(DPN)\}$ *is the set of visible traces starting in* $(M_{init}, init)$ *and ending in* (M_{final}, s) *for some data marking s.*

Definition 16.7 (Perfectly Fitting with Data) *A trace* $\sigma \in (\mathcal{U}_A \times \mathcal{U}_{VM} \times \mathcal{U}_{VM})^*$ *is perfectly fitting* $DPN \in \mathcal{U}_{DPN}$ *if* $\sigma \in \phi(DPN)$. *An event log* $L \in \mathcal{B}((\mathcal{U}_A \times \mathcal{U}_{VM} \times \mathcal{U}_{VM})^*)$ *is perfectly fitting DPN if all of its traces are perfectly fitting.*

Later, we will need to project binding sequences and traces onto subsets of transitions/activities and variables. Therefore, we introduce a generic projection operator $\Pi_{Y,V}(\sigma)$ that removes transitions/activities not in Y and variables not in V.

Definition 16.8 (Projection) *Let* X *be a set of transitions or activities (i.e.,* $X \subseteq T$ *or* $X \subseteq \mathcal{U}_A$). *Let* $Y \subseteq X$ *be a subset and* $V \subseteq \mathcal{U}_{VN}$ *a subset of variable names. Let* $\sigma \in (X \times \mathcal{U}_{VM} \times \mathcal{U}_{VM})^*$ *be a binding sequence or a trace with data.* $\Pi_{Y,V}(\sigma) \in (Y \times (V \nrightarrow \mathcal{U}_{VV}) \times (V \nrightarrow \mathcal{U}_{VV}))^*$ *is the projection of* σ *onto transitions/activities* Y *and variables* V. *Bindings/events unrelated to transitions/activities in* Y *are removed completely. Moreover, for the remaining bindings/events all read and write variables* *not in V are removed.* $\Pi_{Y,V}(L) = [\Pi_{Y,V}(\sigma) \mid \sigma \in L]$ *lifts the projection operator to the level of logs.*

16.2.3 Data Alignments

In this section we extend the alignment theory presented in Chapters 9 and 12 to incorporate the notion of data. Alignments shows how the event log can be replayed on the process model, and they are composed by sequences of moves:

Definition 16.9 (Legal alignment moves) *Let* $DPN = (SN, V, val, init, read, write, guard)$ *be a Petri net with Data, with* $SN = (PN, M_{init}, M_{final})$ *and* $PN = (P, T, F, l)$. *Let* $S_L = \mathcal{U}_A \times \mathcal{U}_{VM} \times \mathcal{U}_{VM}$ *be the universe of events. Let* $S_{DPN} = T \times \mathcal{U}_{VM} \times \mathcal{U}_{VM}$ *be the universe of bindings of DPN. Let be* $S_{DPN}^{\gg} = S_{DPN} \cup \{\gg\}$ *and* $S_L^{\gg} = S_L \cup \{\gg\}$.

A legal move in an alignment *is represented by a pair* $(s_L, s_M) \in (S_L^{\gg} \times S_{DPN}^{\gg}) \setminus \{(\gg, \gg)\}$ *such that*

- (s_L, s_M) *is a move in log if* $s_L \in S_L$ *and* $s_M = \gg$,
- (s_L, s_M) *is a move in model if* $s_L = \gg$ *and* $s_M \in S_{DPN}$,
- (s_L, s_M) *is a move in both without incorrect read/write operations if* $s_M = (t, r, w) \in S_{DPN}$ *and* $s_L = (l(t), r, w) \in S_L$,
- (s_L, s_M) *is a move in both with incorrect read/write operations if* $s_M = (t, r, w) \in S_{DPN}$ *and* $s_L = (l(t), r', w') \in S_L$, *and* $r \neq r'$ *or* $w \neq w'$.

All other moves are considered as illegal.

Definition 16.10 (Data Alignments) *Let* $DPN = (SN, V, val, init, read, write, guard)$ *be a Petri net with Data and* $\sigma \in (S_L)^*$ *be an event-log trace. Let* $\mathcal{A}_{DPN}$ *be the set of legal moves for DPN. A complete alignment of* σ_L *and DPN is a sequence* $\gamma \in \mathcal{A}_{DPN}^*$ *such that, ignoring all occurrences of* $\gg$, *the projection on the first element yields* σ_L *and the projection on the second yields a* $\sigma_P \in \phi_f(DPN)$.

Table 16.2 shows two complete alignments of the process model in Figure 16.2 and the log trace σ_{ex} from Section 1.

As it is explained in Chapter 9, in order to define the severity of a deviation, we introduce a cost function on legal moves: $\kappa \in \mathcal{A}_{DPN} \rightarrow \mathbb{R}_0^+$. This cost function can be used to favor one type of explanation for deviations over others. The cost of each legal move depends on the specific model and process domain and, hence, the cost function κ needs to be defined specifically for each setting. The cost of an alignment γ is the sum of the cost of all individual moves composing it: $\mathcal{K}(\gamma) = \sum_{(s_L, s_M) \in \gamma} \kappa(s_L, s_M)$.

However, we do not aim to find just any complete alignment. Our goal is to find a complete alignment of σ_L and *DPN* which minimizes the cost: an optimal alignment. Let $\Gamma_{\sigma_L, N}$ be the (infinite)set of all complete alignments of σ_L and *DPN*. The alignment $\gamma \in \Gamma_{\sigma_L, DPN}$ is an *optimal alignment* if, for all $\gamma' \in \Gamma_{\sigma_L, N}$, $\mathcal{K}(\gamma) \leq$

Event-Log Trace	Process		Event-Log Trace	Process
(a, {(A,4000)})	(a, {(A,4000)})		(a, {(A,4000)})	(a, {(A,5100)})
(b, {(I,450),(V,false)})	(b, {(I,450),(V,true)})		(b, {(I,450),(V,false)})	(b, {(I,511),(V,true)})
(c, {(D,true)})	(c, {(D,true)})		(c, {(D,true)})	(c, {(D,true)})
(e, ∅)	(e, ∅)		(e, ∅)	(e, ∅)
(f, {(A,5000)})	(f, {(A,3000)})		(f, {(A,5000)})	(f, {(A,5000)})
(b, {(I,450),(V,false)})	(b, {(I,450),(V,false)})		(b, {(I,450),(V,false)})	(b, {(I,511),(V,false)})
(d, {(D,false)})	(d, {(D,false)})		(d, {(D,false)})	(d, {(D,false)})
(e, ∅)	(e, ∅)		(e, ∅)	(e, ∅)
(h, ∅)	≫		(h, ∅)	≫
≫	(g, ∅)		≫	(g, ∅)
(a)			(b)	

Table 16.2: Examples of complete alignments of $\sigma_{example}$ and N. For readability, the read operations are omitted. Of course, read operations for any variable must match the most recent value for that variable. Any move is highlighted with a gray color if it contains deviations, i.e. it is not a move in both without incorrect read/write operations.

$\mathcal{K}(\gamma)$. Note that an optimal alignment does not need to be unique, i.e. multiple complete alignments with the same minimal cost may exist.

Let us consider again our example introduced above. Let us assume to have a cost function κ^s such that $\kappa^s(s_L, s_M) = 1$ if (s_L, s_M) is a visible move in process or a move in log (i.e. $s_L = \gg$ and s_M corresponds to a labeled transition or, conversely, $s_M = \gg$, respectively) or a move in both with incorrect read/write operations and $\kappa^s(s_L, s_M) = 0$ in case of move in both without incorrect read/write operations or a move in model corresponding to an unlabeled transition. The alignment in Table 16.2a has a cost of 6 whereas the alignment in Table 16.2b has a cost 8.[2] It follows that the former is a better alignment. As a matter of fact, it is also an optimal alignment, although it is not the only one. For instance, any variation of such an alignment where the move for f is of the form (now including read operations) $((\mathbf{f}, \{(A, 4000)\}, \{(A, 5000)\})$ $(\mathbf{f}, \{(A, 4000)\}, \{(A, x)\})\})$ with $2250 < x < 4000$ corresponds to an optimal alignment, as well.

As we have mentioned, the data-aware conformance checking is undecidable in the general case. This is caused by the fact that Petri nets with Data are Turing-complete. Therefore, it is not decidable to verify whether a sequence of valid bindings exists that takes from the initial marking to any final marking (M_{final}, s). As a consequence, for instance, it is not possible to find an alignment of a Petri net with Data and the empty log trace. However, the problem becomes decidable (with an exponential complexity) if guards are restricted to linear (in)equalities.

---

[2] They also include a cost of two that is accounted for incorrect read operations, not shown in the alignments, which are caused by incorrect write operations.

16.3 Summary

This chapter presented the concepts of conformance and alignments applied to data-aware process models. This is a time-consuming analysis. Therefore, in the next chapter we present an extension of these concepts to be applied in a decomposed manner.

Chapter 17
Decomposing Data-aware Conformance

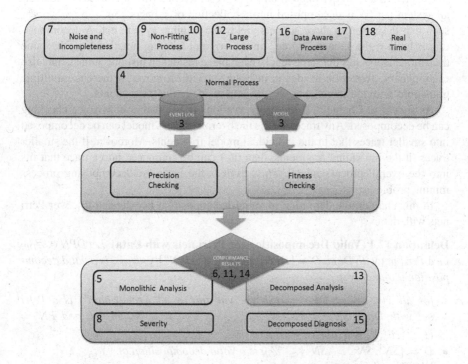

The previous chapter focused on the use of alignments to analyze the data perspective of processes. This chapter presents a decomposition approach of that technique to check conformance of data-aware processes. In particular, the chapter presents a valid decomposition of data-aware models, and a decomposition strategy based on single entry single exit components.

© Springer International Publishing AG 2016
J. Munoz-Gama: Conf. Check. ... in Process Mining, LNBIP 270, pp. 173–179, 2016.
DOI: 10.1007/978-3-319-49451-7_17

17.1 Introduction

In the previous chapter we introduced the concepts of conformance and alignments for data-aware processes, i.e., processes with both control-flow and data
perspectives combined. Checking conformance on data-aware processes is a time-
consuming task. In this chapter we propose a decomposition version of the data-
aware conformance checking in order to reduce the computation time and improve
the understanding of the conformance errors, similar to the one presented in Chapter
13 for control-flow models.

17.2 Valid Decomposition of Data-aware Models

In Chapter 13 a *valid decomposition* [9] is presented in terms of Petri nets: the overall system net SN is decomposed into a collection of subnets $\{SN^1, SN^2, \ldots, SN^n\}$
such that the union of these subnets yields the original system net. A decomposition is *valid* if the subnets "agree" on the original labeling function (i.e., the same
transition always has the same label), each place resides in just one subnet, and also
each invisible transition resides in just one subnet. Moreover, if there are multiple
transitions with the same label, they should reside in the same subnet.

As shown in Chapter 12, these observations imply that conformance checking
can be decomposed. Any trace that fits the overall process model can be decomposed
into smaller traces that fit the individual model fragments. Moreover, if the smaller
traces fit the individual fragments, then they can be composed into a trace that fits
into the overall process model. This result is the basis for decomposing process
mining problems.

In this chapter, the definition of valid decomposition is extended to cover Petri
nets with data.

Definition 17.1 (Valid Decomposition for Petri nets with Data) *Let $DPN \in \mathcal{U}_{DPN}$
be a Petri net with Data. $D = \{DPN^1, DPN^2, \ldots, DPN^n\} \subseteq \mathcal{U}_{DPN}$ is a valid decomposition if and only if:*

- *for all $1 \leq i \leq n$: $DPN^i = (SN^i, V^i, val^i, init^i, read^i, write^i, guard^i)$ is a Petri
  net with Data, $SN^i = (PN^i, M^i_{init}, M^i_{final}) \in \mathcal{U}_{SN}$ is a system net, and $PN^i =
  (P^i, T^i, F^i, l^i)$ is a labeled Petri net,*
- *$D' = \{SN^1, SN^2, \ldots, SN^n\} \subseteq \mathcal{U}_{SN}$ is a valid decomposition of $\bigcup_{1 \leq i \leq n} SN^i$,*
- *$V^i \cap V^j = \emptyset$ for $1 \leq i < j \leq n$,*
- *$DPN = \bigcup_{1 \leq i \leq n} DPN^i$.*

$\mathcal{D}(DPN)$ is the set of all valid decompositions of DPN.

Each variable appears in precisely one of the subnets. Therefore, $V^i \cap V^j = \emptyset$
implies that there cannot be two fragments that read and or write the same data
variables: $\bigcup_{t \in T^i} read^i(t) \cup write^i(t) \cap \bigcup_{t \in T^j} read^j(t) \cup write^j(t) = \emptyset$ for $1 \leq i < j \leq
n$. Moreover, two guards in different fragments cannot refer to the same variable. If

a transition t appears in multiple fragments, then it needs to have a visible unique label. Such a uniquely labeled transition t shared among fragments, may use, read, or write different variables in different fragments. Since $DPN = \bigcup_{1 \leq i \leq n} DPN^i$, we know that, for all t in DPN, $guard(t)$ is the product of all $guard^i(t)$ such that $t \in T^i$. Without loss of generality we can assume that the first k fragments share t. Hence, $guard(t) = guard^1(t) \cdot \ldots \cdot guard^k(t)$. Hence, in a valid decomposition, the guard of a shared transition can only be split if the different parts do not depend on one another. Notice that, the splitting of the data variables is limited by how the variables are used throughout the process, existing a worst-case where all the data variables are used in all the steps of the process.

Based on these observations, we prove that we can decompose conformance checking also for Petri nets with data.

Theorem 17.2 (Conformance Checking With Data Can be Decomposed) *Let $L \in \mathscr{B}((\mathscr{U}_A \times \mathscr{U}_{VM} \times \mathscr{U}_{VM})^*)$ be an event log with information about reads and writes and let $DPN \in \mathscr{U}_{DPN}$ be a Petri net with Data. For any valid decomposition $D = \{DPN^1, DPN^2, \ldots, DPN^n\} \subseteq \mathscr{U}_{DPN}$: L is perfectly fitting Petri net with Data DPN if and only if for all $1 \leq i \leq n$: $\Pi_{A_v(SN^i), V^i}(L)$ is perfectly fitting DPN^i.*

Proof. Let $DPN = (SN, V, val, init, read, write, guard)$ be a Petri net with Data with $SN = (PN, M_{init}, M_{final})$ and $PN = (P, T, F, l)$. Let $D = \{DPN^1, DPN^2, \ldots DPN^n\}$ be a valid decomposition of DPN with $DPN^i = (SN^i, V^i, val^i, init^i, read^i, write^i, guard^i)$, $SN^i = (PN^i, M_{init}^i, M_{final}^i) \in \mathscr{U}_{SN}$, and $PN^i = (P^i, T^i, F^i, l^i)$.

($\Rightarrow$) Let $\sigma_v \in L$ be such that there exists a data marking s such that $(DPN, (M_{init}, init))[\sigma_v \triangleright (DPN, (M_{final}, s))$. This implies that there exists a corresponding σ_b with $(DPN, (M_{init}, init))[\sigma_b\rangle(DPN, (M_{final}, s))$ and $l(\sigma_b) = \sigma_v$. For all $1 \leq i \leq n$, we need to prove that there is a σ_b^i with $(DPN^i, (M_{init}^i, init^i))[\sigma_b^i\rangle(DPN^i, (M_{final}^i, s^i))$ for some s^i. This follows trivially because DPN^i can mimic any move of DPN with respect to transitions T^i: just take $\sigma_b^i = \Pi_{T^i, V^i}(\sigma_b)$. Note that guards can only become weaker by projection.

($\Leftarrow$) Let $\sigma_v \in L$. For all $1 \leq i \leq n$, let σ_b^i be such that $(DPN^i, (M_{init}^i, init^i))[\sigma_b^i\rangle (DPN^i, (M_{final}^i, s^i))$ and $l^i(\sigma_b^i) = \Pi_{A_v(SN^i), V^i}(\sigma_v)$. The different σ_b^i sequences can be stitched together into an overall σ_b s.t. $(DPN, (M_{init}, init))[\sigma_b\rangle(DPN, (M_{final}, s))$ with $s = s^1 \oplus s^2 \oplus \ldots \oplus s^n$. This is possible because transitions in one subnet can only influence other subnets through unique visible transitions and these can only move synchronously as defined by σ_v. Moreover, guards can only be split in independent parts (see Definition 17.1). Suppose that t appears in T_i and T_j, then $guard(t) = guard^i(t) \cdot guard^j(t)$. Hence, a read/write in subnet i cannot limit a read/write in subnet j. Therefore, we can construct σ_b and $l(\sigma_b) = \sigma_v$.

17.3 SESE-based Strategy for a Valid Decomposition

In this section we present a concrete strategy to instantiate the valid decomposition definition over a Petri net with data presented in the previous section (cf. Definition 17.1). Similar to Chapter 12, the proposed strategy decomposes the Petri net with data in a number of Single-Entry Single-Exit (SESE) components, creating meaningful fragments of a process model [69, 64]. SESE decomposition is indicated for well-structured models, whereas for unstructured models some automatic transformation techniques can be considered as a pre-processing step [43].

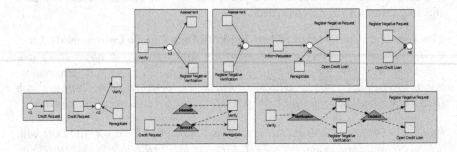

Fig. 17.1: SESE-based decomposition for the running example, with 2-decomposition.

To extend the SESE decomposition strategy presented in Chapter 12 to also account for data, one simply considers its application over the *data workflow graph*, an extension of the workflow graph where the variables and read/write arcs are also included.

Definition 17.3 (Data Workflow Graph) *The* data workflow graph *of a Petri net with Data* $(((P, T, F, l), M_{init}, M_{final}), V, val, init, read, write, guard)$ *with data arcs* $R = \{(v, t) | v \in read(t)\}$ *and* $W = \{(t, v) | v \in write(t)\}$ *is the workflow graph* $DWG = (S, E)$ *with* $S = P \cup T \cup V$ *and* $E = F \cup R \cup W$.

The SESE decomposition proposed to analyze the conformance of Petri nets with data, is similar to the one presented in Chapter 12 but considering data workflow graph instead. Algorithm 17.1 describes the steps necessary to construct a SESE decomposition. The arcs are partitioned in SESEs by means of creating the RPST from the data workflow graph, and selecting a particular set of SESES over it. Once the partitioning is done, a subnet is created for each part. Subnets contradicting some of the requirements of Definition 17.1 (e.g. sharing places, invisible or duplicate transitions, variables, or transitions with non-splitting guards) are merged to preserve the valid decomposition definition.

Figure 17.1 shows the decomposition for the example of Figure 16.2, where the RPST is partitioned using the 2-decomposition algorithm (cf. Chapter 12), i.e.,

---

Algorithm 17.1 SESE-based Decomposition

1: Build data workflow graph *DWG* from *F*, *R*, *W*
2: Compute *RPST* from *DWG*
3: Compute SESE decomposition *D* from the *RPST*
4: Compute and merge subnets if necessary to preserve valid decomposition.
5: **return** valid decomposition where perspectives are decomposed altogether

---

SESEs of at most 2 arcs[1]. To ensure a valid decomposition is obtained, step 4 of Algorithm 17.1 combines multiple SESE fragments into larger fragments, which are not necessarily SESEs anymore.

17.4 Implementation and Experimental Results

The approach requires a Petri Net with Data and an event log as input and returns as many bags of alignments as the number of fragments in which the Petri Net with Data has been decomposed. Each bag refers to a different fragment and shows the alignments of each log trace and that fragment. A second type of output is also produced in which the alignments' information is projected onto the Petri net with Data. Transitions are colored according to the number of deviations: if no deviation occurs for a given transition, the respective box in the model is white-colored. The filling color of a box shades towards red as a larger fraction of deviations occur for the corresponding transition. Something similar is also done for variables: the more incorrect read/write operations occur for a variable, the more the variable is shown with a color close to red. This output is extremely interesting from an end-user viewpoint as it allows for gaining a helicopter view on the main causes of deviations [50].

The approach has been evaluated using a number of synthetic event logs and also a real-life process. The approach has been evaluated using the model in Figure 16.2 and with a number of event logs that were artificially generated. In particular, we have generated different event logs with the same number of traces, 5000, but increasing number of events, meaning that, on average, traces were of different length. To simulate that, for each simulated process execution, an increasing number of renegotiations was enforced to happen. Traces were also generated so as to contain a number of deviations: the event logs were generated in a way that 25% of transitions fired violating the guards.

Figure 17.2 shows the results of checking for conformance of the different event logs and the process model, comparing the SESE-based decomposition with $k = 2$ with the case in which no decomposition is made. To check the conformance of each fragment, we used the technique reported in [50]. Each dot in the chart indicates a different event log with traces of different size. The computation time refers to the

---

[1] Although the SESEs have at most two arcs, this is not guaranteed for the final subnets, i.e., some subnets are merged to preserve the valid decomposition definition.

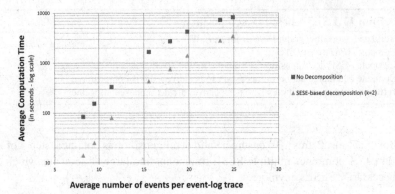

Fig. 17.2: Computation time for checking the conformance of the Petri net with Data in Figure 16.2 and event logs of different size. The Y axis is on a logarithmic scale.

conformance checking of the whole event logs (i.e., 5000 traces). The decomposed net is the same as in Figure 17.1. Regarding the cost function, we assign cost 1 to any deviation; however, this could be customized based on domain knowledge. The results show that, for every combination of event log and process model, the decomposition significantly reduces the computation time and the improvement is exponential in the size of the event log.

To assess the practical relevant of the approach, we also performed an evaluation with a Dutch financial institute. The process model was provided by a process analyst of the institute and consists of 21 transitions: 13 transitions with unique labels, 3 activities labels shared between 2 transitions (i.e. 6 transitions in total), plus 3 invisible transitions. The model contains twelve process variables, which are read and written by the activities when being executed. We were also provided with an event log that recorded the execution of 111 real instances of such a process; overall, the 111 log traces contained 3285 events, which means roughly 29.6 events per trace. We checked the conformance of this process model and this event log, comparing the results when the model has or has not been decomposed in small fragments. For conformance checking, here we used the technique reported in [53] since the provided process model breaks the soundness assumptions required by [50]. For this experiment round, the additional optimizations proposed in [53] were deactivated to allow for a fair comparison.

The application of the decomposition approach to this real-life case study has shown tremendous results: the conformance checking has required 52.94 seconds when the process model was decomposed using the SESE-based technique presented in Section 17.3; conversely, it required 52891 seconds when the model was not decomposed. This indicates that decomposing the process model allowed us to save 99.999% of the computation time. As a matter of fact, we tried for different values of SESE parameter k but we obtained similar results: the computation time did not move away for more than 1 second. The reason of this is related to the fact that every decomposition for any value of k always contained a certain fragment, along with others. Indeed, that fragment could not be decomposed any further than a given extent. Since the computation time was mostly due to constructing alignments with that fragment, no significant difference in computation time could be observed when varying k.

17.5 Summary

Checking conformance between an event log and a process model is known to be a complex procedure. That complexity explodes even more when we consider data-aware conformance checking with multi-perspective models. This chapter proposed the extension of control-flow decomposed conformance checking techniques to alleviate the computation time of data-aware processes. The decomposition definition proposed is proven to be correct from a conformance point of view, and the experiments showed a significant reduction in time with respect to the monolithic approach. In the future, new decomposition strategies may be considered, and the proposed techniques may be extended to other conformance dimensions, such as precision.

Chapter 18
Event-based Real-time Decomposed Conformance Checking

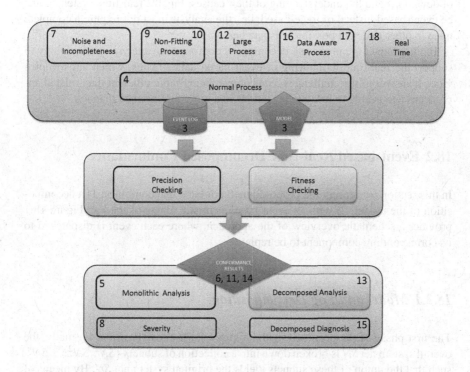

Previous chapters illustrate the use of decomposition to analyze fitness issues of the processes. In this chapter we present an approach to analyze conformance checking in a real-time setting. This approach is based on the decomposition based on single entry single exit components proposed in previous chapters.

© Springer International Publishing AG 2016
J. Munoz-Gama: Conf. Check. ... in Process Mining, LNBIP 270, pp. 181–188, 2016.
DOI: 10.1007/978-3-319-49451-7_18

18.1 Introduction

In this chapter we use the application of decomposed techniques to check conformance in event-based real-time systems. Unlike forensic conformance checking, where conformance is analyzed a posteriori once the case has finished, real-time techniques check conformance on the fly. Real-time checking techniques are specially indicated for monitoring, immediate fraud detection and governance, risk and compliance verification, and failure protection.

Although both forensic and real-time checking analyze the conformance between models and reality, the latter presents a list of new challenges for its applicability. First, the proposed approach should consider the computation time a priority. The analysis must be conducted on a regular basis and the results must be output in a short period of time. Second, given the periodicity of the analysis and the urge for conclusions, real-time approaches must focus on the fine-grained localization of deviations and the understanding of their causes. Finally, real-time systems must be event-based instead of trace-based, i.e., the analysis must not require a complete trace in other to detect possible mismatches.

In this chapter we propose a decomposed conformance analysis methodology to support the real-time monitoring of event-based data streams, which aims to provide an answer to the challenges listed above, being more efficient than related approaches and is able to localize deviations in a more fine-grained manner.

18.2 Event-based Real-time Decomposed Conformance

In this section we present the proposed methodology, that combines: 1) a decomposition of the model, 2) with an event-based heuristic replay of the log. Figure 18.1 provides a schematic overview of the approach, where each event is dispatched to its corresponding component to be replayed.

18.2.1 Model and Log Decomposition

The first phase of the proposed methodology entails *decomposition*. Formally, the overall system net SN is broken down into a collection of subnets $\{SN^1, SN^2, \ldots SN^n\}$ such that the union of these subnets yields the original system net SN. By means of decomposing the original model into a set of subnets we aim to achieve the following goals. First, fragment the conformance problems into a set of more comprehensive semantic elements to aid the diagnosis. Second, restrict the possible pernicious effects of the heuristics decisions taken during the conformance analysis (see Phase 3 below). Third, speed-up the analysis in comparison to non-decomposed conformance checking techniques.

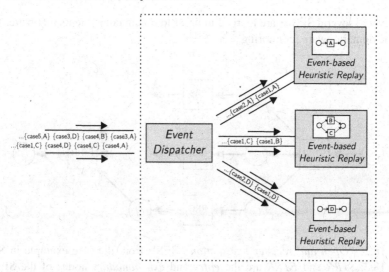

Fig. 18.1: Architectural overview of the developed real-time decomposed conformance checking technique.

Due to the final goal of analyzing conformance, not all possible decomposition approaches are appropriate for this task. Only those *valid* decompositions that preserve the conformance integrity should be considered (cf. Chapter 12). That is, given the original net and the decomposed version, *the original net perfectly conforms iff all the subnets in the decomposed setting perfectly conforms.* In other words, no conformance anomalies should be lost or introduced in the transition from the overall model to the decomposed one. As it is presented in Chapter 12, a valid decomposition—applicable on Petri nets— is defined informally as the decomposition that satisfies the following conditions:

1. Each arc of the overall net belongs to exactly one of the subnets.
2. Each place of the overall net belongs to exactly one of the subnets.
3. Invisible transitions appears in precisely one of the subnets.
4. Visible, duplicate transitions appear in precisely one of the subnets.
5. Visible, non-duplicate transitions may appear in multiple subnet.

In other words, all elements in the original Petri net model must belong to a subnet, but only visible, non-duplicate transitions can be shared among several subnet.

As it is mentioned in previous chapters, there exist several possible valid decomposition strategies: minimal, passages, SESEs, etc. In this methodology we consider a valid decomposition based on the SESE decomposition proposed in Chapter 12, i.e. subgraphs in the workflow graph defined over a system net having single entry and exit boundary nodes. SESEs perfectly reflect the idea of subprocesses within the main process, important to obtain a meaningful real-time decomposition. Figure 18.2 shows an example of SESE for the illustrative case shown in Section 18.3. The SESE decomposition can be combined with a user-supervised post-processing step where several SESEs are merged in order to obtain components that better fulfill the domain-aware monitoring.

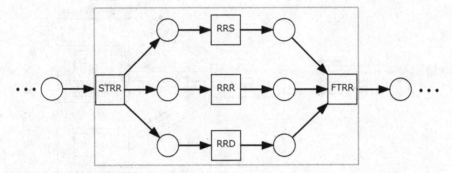

Fig. 18.2: *"Open and register transaction"* SESE from the case example in Section 18.3. *STRR* and *FTRR* are the *entry* and *exit boundary* nodes of the SESE, respectively. The rest of places and transitions are *interior* nodes of the SESE.

Once a system net has been decomposed into a set of submodels, this collection of models is passed to a central *event dispatcher*, which also serves to listen for incoming events. For each submodels, it is examined whether it contains a transition t which maps to the incoming event e. If it does, this indicates that the event at hand should be replayed on this particular submodel (multiple such submodels can be found), and the event is passed forward to this model fragment.

18.2.2 Event-based Heuristic Replay

Once it is determined which process model fragment(s) should parse the incoming event, the actual *replay* of this event on each such fragment is performed. In previous chapters – Chapters 9 and 12 – we illustrate the use and benefits of a optimal conformance checking based on alignments. However, given the event-based nature of the methodology, and the need for efficient approaches in real-time systems, in this chapter we propose the use of replay based conformance techniques. In the seminal work [77], a "fitness" metric is presented to describe the extent to which event traces can be associated with valid execution paths in the process model, and an "appropriateness" metric is proposed to assess whether the process model describes the observed behavior accurately enough. The aforementioned approach *replays* the traces of the log in the model to evaluate these metrics. One of the drawbacks of this approach is that for *undeterministic* models, the heuristics used in the replay may lead to overestimating the metrics, due to the artificial creation of superfluous tokens in the model. Several solutions have been proposed to overcome this issue. Weidlich et al. propose a system to check process model consistency based on "behavioral profiles" [91, 83]—which can be derived in a straightforward and efficient manner but with loss of some granularity regarding the exact traces which can be accepted by the model at hand.

In this chapter we propose the use of a replay algorithm based on the work of vanden Broucke et al. [28, 29]. The informal idea is the following: for each process model fragment, a state list is maintained denoting the current marking reached by the currently-running process instances. When an event is queued for replay by a process fragment, the state linked to process instance is progressed by investigating whether there exists an enabled transition for such activity. The outcome of this evaluation determines if the process model is showing discrepancies or not.

Some additional remarks should be provided at this point. First of all, we note that we apply a heuristic, event-granular replayer similar to the one applied in [28]. The reasoning behind the choice to opt for a replayer playing the token game instead of an alternative approach such as alignment or behavioral profile based techniques [18, 83, 91] are twofold. First, alignment and behavioral profile based replayers perform their analysis on a trace, rather than event level, meaning that a complete process instance needs to be finalized in order to align the log trace with a process model transition sequence. As we are dealing with event streams which need to be analyzed in a real-time manner, an event-granular replay strategy is required. Second, alternative approaches suffer from scalability issues which make them unsuitable in a real-time context.

A second remark entails the way in which decision points are resolved by the replayer. Put briefly, whenever multiple (enabled) transitions are mapped to the same event log activity within a process model and/or whenever multiple invisible activities are enabled, the replayer needs to determine which transition to execute to handle the activity at hand. Note that—in extreme edge cases—it is possible that the forced firing of a non-enabled transition should be preferred if this avoids several other violations later in the event trace [85]. A replay strategy is put forward which prefers the firing of enabled transition mapped to the activity at hand first, followed by the set of silent transitions, followed by the set of non-enabled transition mapped to the activity at hand. If the chosen set contains multiple transition candidates, a one-step look-ahead procedure is executed to determine which candidate enables the execution of the following activity (if no such candidate can be found, a random one is chosen). For the multitude of process models, this look-ahead suffices to resolve any ambiguities. However, since we are dealing with streaming event data in this context, we possess no knowledge about the events that will arrive in the future, preventing the execution of the look-ahead procedure. There are three proposed strategies to deal with this issue. First, disabling the look-ahead altogether and assuming that the model is deterministic enough to handle incoming events without taking the context into account. Second (another extreme), restarting the replay of the full trace each time an event is added, thus allowing the replayer to revise earlier decisions. Note however that the replayer is configured such that no new violations may be introduced related to historical activities. In practice, this means that the replayer can revise the state chain by modifying the execution of silent transitions, selecting alternative albeit *also enabled* transition mapped to a particular activity for activities which were parsed correctly, or selecting alternative disabled transition, although only for activities which were not parsed correctly. The third method combines these two extremes by considering a part of the executed transition sequence as "frozen", only allowing revisions for the last n steps.

As a third remark, as it is aforementioned, one of the drawbacks of "token game"-based replayers entails the possible creation of superfluous tokens, enabling subsequently for too much behavior. However, as was mentioned, the decomposition of a process model restricts the possible pernicious effects of the heuristics decisions taken during the conformance analysis, as each model is now limited to dealing with a smaller subset of behavior. In addition, as superfluous tokens are created following the forced firing of violating activities, the process instance or model fragment at hand is likely to be immediately indicated as "dubious" at this point, lowering the trustfulness of following events within this instance of model fragment, independent of the replay strategy being applied.

The results of the replay analysis can be reported and visualized. Remark that, naturally, these actions can be performed while the actual conformance analysis is running. In general, two ways of result follow-up are supported by our architecture. The first one consists of the logging of various statistics by the running worker threads and replayers, which is polled regularly by decoupled components (e.g. a real-time dashboard or perhaps logged to a persistent data store). The second manner by which results can be interpreted consists of the definitions of various triggers which are to be fired once certain criteria are met, such as a model fragment overshooting a certain error rate threshold, for instance, of a high-risk activity or model fragment being violated. The actions which can be undertaken as a result are self-explanatory, e.g. sending warnings, or halting running process instances or even the complete system.

18.3 Experimental Results

To benchmark the performance of our developed real-time conformance analysis technique against related approaches, a fitting event log was generated (based on the model depicted in Figure 14.2) containing ten thousand process instances (678864 events). A non-conforming ("noisy") variant of this event log was produced by inducing noise (inserting, deleting, and swapping of events) so that 10% of the included events are erroneous.

We compare our proposed technique against the alignment based replay technique by Adriansyah et al. [18] as well with the original implementation of the token-game based heuristic replayer [28]. Both the non-decomposed and decomposed variants of these techniques were included.

Figure 18.3 depicts the performance results of the experiment, showing the amount of time taken (x-axis) to check the conformance of the included event logs (the y-axis represents the cumulative ratio of event checks performed). As can be seen, our proposed real-time conformance analysis technique performs

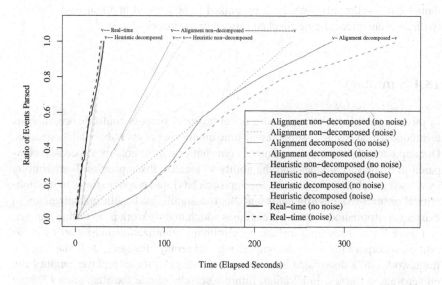

Fig. 18.3: Comparison of replay performance for the included techniques in the experimental setup, showing the time taken per technique to replay the given event log.

competitively with respect to related techniques. During the experimental run, a maximum throughput rate (number of events checked per second) was reached at 35000 with the experiment running on a single consumer laptop with three worker threads. Some additional remarks should be provided however when interpreting Figure 18.3. First, note that our proposed technique performs similarly as the heuristic decomposed replay technique, but note that proposed technique executes a *conformance check on an event-granular basis* and thus can be applied in a real-time monitoring setting, whereas the other techniques do so on a trace-granular level (i.e. a complete trace should be provided to perform the replay procedure). However, the event log is of sufficient size so that a step-wise effect is not apparent in Figure 18.3. Second, the replay procedure of the existing techniques was modified such that *each trace is checked independently* of the log context, meaning that no distinct trace grouping is performed over the log and that each trace is checked as if it belonged to an event log that only contains the particular trace, so as to better assess the performance of these techniques in a real-time scenario (where the complete trace and log are unknown as events are arriving), rather than a post-hoc scenario where the complete event log is provided as-is. Note that—for the alignment based technique—this causes the non-decomposed version to perform better than the decomposed one. This is perhaps an unexpected result, but it is caused by the fact that the alignment based techniques are geared towards checking—and as such expect—event logs as a whole. We thus emphasize the fact that these techniques have—currently—not been optimized to be applied in a real-time scenario (with an event stream being checked instead of an historical log).

18.4 Summary

In this chapter we have presented a novel business process conformance analysis technique which is able to support real-time monitoring of event-based data streams. Our approach offers a number of novel contributions, most notably a speed-up compared to related techniques and the ability to localize discrepancies. Furthermore, by allowing real-time monitoring, the approach has rapid response times in mission-critical or high-risk environments and this is a significant benefit in comparison to existing conformance checking techniques which mainly work in an offline manner.

Future lines of research include: streamlining visualization and reporting capabilities, incorporating other decomposition and replay strategies, and adapting the framework into a distributed implementation, where different replayer engines run on separate machines. In addition, future research include the adaptation of these techniques to a strict steaming scenario, where the cases have no identification.

Part IV
Conclusions and Future Work

Chapter 19
Conclusions

This chapter concludes the book, summarizing the contributions, and providing a final reflection on the challenges and future directions of conformance checking.

19.1 Conclusion and Reflection

In the early days of process mining, most of the research in process mining was mainly focused on discovery techniques, neglecting the importance of conformance. Several discovery approaches were presented, using a wide range of different techniques. However, measuring the adequacy of the provided results was never a priority, and only simple measures were used, making it difficult to determine the most adequate approach for each case. Moreover, most analysis focused on verifying the properties of the model (e.g., has the model deadlocks?), or the properties of the log (e.g., is activity x and y always executed by a different person?), instead of evaluating whether the model represented faithfully the reality observed in the log, i.e., conformance.

However, in the recent years this tendency has finally changed, and more research effort has been focused on the conformance dimensions. Since the thesis [73], several master and doctoral thesis has focused, totally or partially, on conformance. Thesis like [17], [29], [32], or [58], represent a good example of that effort. In addition, several papers were presented in journals, conferences and workshops to analyze conformance: works addressing all conformance dimensions, such as [18, 9] for fitness, [59, 21, 28] for precision, [28, 10] for generalization, or [35] for simplicity, to enumerate some examples; works addressing conformance using replay-based strategies such as [28], and also align-based strategies such as [18]; works addressing conformance in a decomposed way, such as [8, 9, 62]; and works addressing conformance for multi-perspective models such as [50]. New benchmarking frameworks for conformance analysis has been presented, such as [31], incorporating and using these new approaches, for example in [89], to evaluate the quality of state-of-the-art discovery algorithms.

© Springer International Publishing AG 2016
J. Munoz-Gama: Conf. Check. ... in Process Mining, LNBIP 270, pp. 191–195, 2016.
DOI: 10.1007/978-3-319-49451-7_19

19.2 Summary of Contributions

The main theme of this book is the conformance checking and diagnosis between behavior observed in the log and process models. The contributions of the book can be divided in two groups according to their purposes: techniques to evaluate the *precision* dimension, and techniques to decompose the conformance diagnosis.

- **Precision in Conformance Checking.** The *precision* is the dimension in conformance checking that measures the degree of a process model to describe only the behavior observed in the log, and no more.

  Chapters 4, 5, and 6 studied the limitations of the approaches based on comparing model and log relations, and presented a different approach to study precision. This approach is based on detecting *escaping arcs*, in other words, those points where the model allows more behavior than the one observed in the log. The escaping arcs are weighted and used to provide a metric for the precision dimension. Moreover the escaping arcs, and the situations leading to them, define possible points for a future actuation, in order to achieve a more precise model.

  In Chapter 7, the escaping arcs theory is revisited to incorporate the notion of *robustness*. The refined approach consider the frequency of the observed traces in order to be less affected by infrequent or noisy observed behavior. The chapter also presented the notion of *confidence* over the precision metric provided. A narrow confidence interval indicates a high confidence on the provided metric, while a wide interval denotes a likely possibility that precision metric changes in the future. The confidence interval is determined according to a giving parameter indicating the future to consider, where a low value indicates a close future and a high value a distance future. The bounds of the interval represent the likelihood of new escaping arcs to appear, or disappear, within the considered future. Chapter 8 presented a method to assess the *severity* of the escaping arcs detected. The severity proposed is based in a multi-factor analysis, with factors such as the weight of the escaping arc within the process or the criticality of the action allowed. Escaping arcs with high severity pinpoint imprecise situations that should be addressed urgently.

  Chapters 9, 10, and 11 presented a different approach to analyze the escaping arcs of a system. While the previous proposed technique detects the escaping arcs directly from the log, this different approach uses a pre-processing step, where the observed behavior is *aligned* with the modeled behavior. Aligning observed and modeled behavior can solve situations, such as unfitting and undeterminism, in a global and optimal way, whereas the direct use of the log would require the use of heuristics. However, the aligning of behaviors is a time-consuming operation not suitable for all the scenarios. The chapter presented the use of aligned behavior to derive escaping arcs, including as well situations where the observed traces in the log can be optimally aligned with several model sequences. The chapter also proposed different levels of abstractions on the precision computation, where the order of the activities is not considered, or where the direction in which the log traces are processed is reverted to achieve a more balanced precision.

- **Decomposed Conformance Diagnosis.** Conformance checking is a time-consuming task. Moreover, identifying the causes of the conformance problems is sometimes far from easy. *Decomposition* techniques try to alleviate these two problems by decomposing the processes in parts, and analyzing them separately. In this part, the conformance focus changes from the precision dimension of the previous part to the fitness dimension.

  Chapters 12, 13, and 14 studied the different decomposition methods in process mining, and proposed a decomposition conformance checking based on *Single-Entry Single-Exit (SESE)* components. SESEs are components with a clear interface with the rest of the model, (i.e., one entry node and one exit node), representing subprocesses within the main process. Decomposing a model in SESEs alleviates the conformance checking analysis, while at the same time, the meaningful decomposition provides a better understanding of what subprocesses have conformance problems.

  Chapter 15 extended the diagnosis capacities of the decomposition methods, providing mechanisms to analyze the conformance in a topological way and a hierarchical way. A topology of a decomposition represents the connections between components, and it can be used to detect areas with conformance problems. A hierarchical conformance analysis allows to zoom-in and zoom-out on a hierarchy of components in order to get a better understanding of the situation and the cause of the conformance problems.

  Chapter 16 and 17 proposes a decomposition of multi-perspective models, in order to alleviate the computation time and to aid the conformance diagnosis. In particular, the proposed decomposition focuses on models with control-flow and data perspective, aligning both flow and data in a distributed way.

  Chapter 18 addresses the real-time monitoring of conformance deviations by means of decomposition and event-based heuristic replay. The decomposed setting aims at identifying the subprocesses cause of the conformance anomalies, while the replay adapts the analysis to the event-based nature of real-time monitoring scenarios.

19.3 Challenges and Directions for Future Work

The work presented in this book focuses on the area of conformance checking, a first step to perform conformance analysis which was impossible until now. In this section we list some of the possible future research paths to follow:

- **New metrics, new dimensions.** Each conformance dimension states the property all metrics for that dimension should measure. However, the details on how this dimension is quantified relies on each specific metric, and each metric assumes a different interpretation of the dimension. For example, *PM* [92] defines fitness as the number of correct parsed traces divided by the number of traces in the event log, while the metric f [77] is more fine-grained because it also considers the problems (missing and remaining tokens) that occurred during the log replay. In precision, a'_B [77] derives sometimes follows and precedes relations for the model and the log and compares them, while etc_p [59] bases its estimation on detecting escaping arcs. Each proposed metric has advantages and disadvantages that make it appropriate for particular scenarios. For example, a precision metric such as etc_p, less time consuming than a'_B, is more suited to be incorporated within a genetic algorithm where conformance is executed constantly [35]. Therefore, the study of new conformance metrics is an area that must be explored as future research, extending the base of metrics to be used. Moreover, limiting conformance to only four dimensions (fitness, precision, generalization, and simplicity) may be considered too restrictive, and future research must be open to explore other possible dimensions, such as the completeness, i.e., is the model describing the complete process or only a part?

- **Decomposed alignment of observed and modeled behavior.** As experiments show, aligning observed and modeled behavior can be a time-consuming task [17]. Chapter 12 and other similar works alleviate this cost by means of decomposing the process. However, as it is remarked in the same chapter, the goal of decomposing is not to aligning the whole log and model (i.e., the original problem), but to align at component level to provide localized conformance information. It is proven by counterexample that the simple concatenation of local optimal alignments does not result in a global optimal alignment. However, the use of decomposed techniques have provided equally important results for the global setting. For example, [9] provides a lower bound on the global alignment cost based on local alignments. The use of decomposed techniques is a promising research path to alleviate the cost of the alignments, and lots of efforts has been put in the recent months. That may include, for example, the study of what properties the traces, model, or local alignments must satisfy so the lower bound results in the exact cost, or the development of divide-and-conquer strategies for a bottom-up construction of the alignments.

- **Decomposed conformance for other dimensions.** Chapter 12 shows how to check perfectly fitting traces in a decomposed way. However, conformance checking also includes three other dimensions. The use of decomposed techniques to measure those dimensions is an interesting future research path, but it is also a challenge far from trivial. For example, metrics such as etc_p [59] and a'_B [77] define precision as a global property, and therefore, a direct decomposition is not possible. There are several options that address the issue and can be explored: the real cost of metrics like a_p [21] comes from computing the alignments they rely on, and thus, a faster decomposed alignment of the behaviors will reduce the computation cost of the metric; in other cases, a simple conformance analysis can be done at local level and then can be smartly aggregated at a global level, similar to a *Map Reduce* strategy; a final option is the definition of new metrics that consider a redefinition of the dimension in a non-global way. Similarly, decomposition can be used to measure simplicity, for example, measuring the simplicity of the components and how they are connected.

- **Visualization and diagnosis.** The usefulness of results is directly related with how the results are provided. Results poorly displayed may limit their transmitted information. This is specially important in the conformance checking area, where results tend to be large (e.g., set of escaping arcs, tokens missing, misalignments, ...) and pinpointing future lines of actuation withing the organization. The book provides several mechanisms in that direction (e.g., severity of the escaping arcs, or detection of problematic areas using the topology of a decomposition). However, this is just the tip of the iceberg and more approaches should follow. Some future work possibilities include the aggregation of conformance problems by the underlying cause behind them, or new ways to visually represent those problems.
- **Model repair.** The conformance diagnosis approaches presented in this book indicate points where the model does not adequately describe the reality. These points indicate possible parts of the model to be repaired, but the approaches rely on the user to perform iterative improvements on the model. The use of the conformance results within a fully automated approach to repair the models, similar to [44] and [36], is an interesting topic for further research. This becomes especially interesting in the decomposed setting, where the model can be repaired component by component. Moreover, other scenarios to consider are also possible, such as models that cannot be modified, opening the door to theories like *supervisory control* [71] in order to derive a controller to supervise the model execution.

References

1. Extensible Event Stream (XES). www.xes-standard.org
2. IEEE Task Force on Process Mining – Case Studies . http://www.win.tue.nl/ieeetfpm/doku. php?id=shared:process_mining_case_studies
3. van der Aalst, W.M.P.: Verification of Workflow Nets. In: P. Azéma, G. Balbo (eds.) Application and Theory of Petri Nets, PETRI NETS '97, *Lecture Notes in Computer Science*, vol. 1248, pp. 407–426. Springer (1997)
4. van der Aalst, W.M.P.: The Application of Petri Nets to Workflow Management. Journal of Circuits, Systems, and Computers **8**(1), 21–66 (1998)
5. van der Aalst, W.M.P.: Formalization and verification of event-driven process chains. Information & Software Technology **41**(10), 639–650 (1999)
6. van der Aalst, W.M.P.: Using Process Mining to Generate Accurate and Interactive Business Process Maps. In: W. Abramowicz, D. Flejter (eds.) Business Information Systems Workshops, BIS'09, *Lecture Notes in Business Information Processing*, vol. 37, pp. 1–14. Springer (2009)
7. van der Aalst, W.M.P.: Process Mining - Discovery, Conformance and Enhancement of Business Processes. Springer (2011)
8. van der Aalst, W.M.P.: Decomposing Process Mining Problems Using Passages. In: S. Haddad, L. Pomello (eds.) 33rd International Conference on Application and Theory of Petri Nets, PETRI NETS '12, *Lecture Notes in Computer Science*, vol. 7347, pp. 72–91. Springer (2012)
9. van der Aalst, W.M.P.: Decomposing Petri nets for process mining: A generic approach. Distributed and Parallel Databases **31**(4), 471–507 (2013)
10. van der Aalst, W.M.P., Adriansyah, A., van Dongen, B.F.: Replaying history on process models for conformance checking and performance analysis. Wiley Interdisc. Rew.: Data Mining and Knowledge Discovery **2**(2), 182–192 (2012)
11. van der Aalst, W.M.P., ter Hofstede, A.H.M.: YAWL: yet another workflow language. Inf. Syst. **30**(4), 245–275 (2005)
12. van der Aalst, W.M.P., ter Hofstede, A.H.M., Kiepuszewski, B., Barros, A.P.: Workflow Patterns. Distributed and Parallel Databases **14**(1), 5–51 (2003)
13. van der Aalst, W.M.P., Reijers, H.A., Song, M.: Discovering Social Networks from Event Logs. Computer Supported Cooperative Work **14**(6), 549–593 (2005)
14. van der Aalst, W.M.P., Rubin, V., Verbeek, H.M.W., van Dongen, B.F., Kindler, E., Günther, C.W.: Process mining: a two-step approach to balance between underfitting and overfitting. Software and System Modeling **9**(1), 87–111 (2010)
15. van der Aalst, W.M.P., Weijters, T., Maruster, L.: Workflow Mining: Discovering Process Models from Event Logs. IEEE Trans. Knowl. Data Eng. **16**(9), 1128–1142 (2004)
16. Accorsi, R., Stocker, T.: On the exploitation of process mining for security audits: the conformance checking case. In: S. Ossowski, P. Lecca (eds.) Proceedings of the ACM Symposium

on Applied Computing, SAC 2012, Riva, Trento, Italy, March 26-30, 2012, pp. 1709–1716. ACM (2012). DOI 10.1145/2245276.2232051

17. Adriansyah, A.: Aligning Observed and Modeled Behavior. Ph.D. thesis, Technische Universiteit Eindhoven, Eindhoven, The Netherlands (2014)

18. Adriansyah, A., van Dongen, B.F., van der Aalst, W.M.P.: Conformance Checking Using Cost-Based Fitness Analysis. In: 15th IEEE International Enterprise Distributed Object Computing Conference, EDOC'11, pp. 55–64. IEEE Computer Society (2011)

19. Adriansyah, A., van Dongen, B.F., van der Aalst, W.M.P.: Memory-Efficient Alignment of Observed and Modeled Behavior. Tech. Rep. BPM-03-03, BPMcenter.org (2013)

20. Adriansyah, A., Munoz-Gama, J., Carmona, J., van Dongen, B.F., van der Aalst, W.M.P.: Alignment Based Precision Checking. In: M.L. Rosa, P. Soffer (eds.) 8th International Workshop on Business Process Intelligence, BPI'12, *Lecture Notes in Business Information Processing*, vol. 132, pp. 137–149. Springer (2012)

21. Adriansyah, A., Munoz-Gama, J., Carmona, J., van Dongen, B.F., van der Aalst, W.M.P.: Measuring precision of modeled behavior. Inf. Syst. E-Business Management **12** (2014). (to appear)

22. Adriansyah, A., Sidorova, N., van Dongen, B.F.: Cost-Based Fitness in Conformance Checking. In: B. Caillaud, J. Carmona, K. Hiraishi (eds.) 11th International Conference on Application of Concurrency to System Design, ACSD'11, pp. 57–66. IEEE (2011)

23. Banescu, S., Petkovic, M., Zannone, N.: Measuring Privacy Compliance Using Fitness Metrics. In: A.P. Barros, A. Gal, E. Kindler (eds.) 10th International Conference on Business Process Management, BPM'12, *Lecture Notes in Computer Science*, vol. 7481, pp. 114–119. Springer (2012)

24. Bezerra, F.d.L., Wainer, J.: Algorithms for anomaly detection of traces in logs of process aware information systems. Inf. Syst. **38**(1), 33–44 (2013)

25. Bezerra, F.d.L., Wainer, J., van der Aalst, W.M.P.: Anomaly Detection Using Process Mining. In: T.A. Halpin, J. Krogstie, S. Nurcan, E. Proper, R. Schmidt, P. Soffer, R. Ukor (eds.) 10th International Workshop on Business Process Modeling, Development, and Support, BP-MDS'09, *Lecture Notes in Business Information Processing*, vol. 29, pp. 149–161. Springer (2009)

26. Bose, R.P.J.C., van der Aalst, W.M.P., Zliobaite, I., Pechenizkiy, M.: Handling Concept Drift in Process Mining. In: H. Mouratidis, C. Rolland (eds.) Advanced Information Systems Engineering, CAiSE '11, *Lecture Notes in Computer Science*, vol. 6741, pp. 391–405. Springer (2011)

27. Box, G.E.P., Hunter, W.G., Hunter, J.S.: Statistics for experimenters : an introduction to design, data analysis, and model building. Wiley series in probability and mathematical statistics. J. Wiley & Sons (1978)

28. vanden Broucke, S.K., Weerdt, J.D., Vanthienen, J., Baesens, B.: Determining Process Model Precision and Generalization with Weighted Artificial Negative Events. IEEE Transactions on Knowledge and Data Engineering **99**(PrePrints), 1 (2013). DOI 10.1109/TKDE.2013.130

29. vanden Broucke, S.K.L.M.: Advances in Process Mining: Artificial Negative Events and Other Techniques. Ph.D. thesis, Katholieke Universiteit Leuven, Leuven, Belgium (2014)

30. vanden Broucke, S.K.L.M., Weerdt, J.D., Baesens, B., Vanthienen, J.: Improved Artificial Negative Event Generation to Enhance Process Event Logs. In: J. Ralyté, X. Franch, S. Brinkkemper, S. Wrycza (eds.) 24th International Conference on Advanced Information Systems Engineering, CAiSE'12, *Lecture Notes in Computer Science*, vol. 7328, pp. 254–269. Springer (2012)

31. vanden Broucke, S.K.L.M., Weerdt, J.D., Vanthienen, J., Baesens, B.: A comprehensive benchmarking framework (CoBeFra) for conformance analysis between procedural process models and event logs in ProM. In: IEEE Symposium on Computational Intelligence and Data Mining, CIDM'13, pp. 254–261. IEEE (2013)

32. Buijs, J.C.A.M.: Flexible Evolutionary Algorithms for Mining Structured Process Models. Ph.D. thesis, Technische Universiteit Eindhoven, Eindhoven, The Netherlands (2014)

33. Buijs, J.C.A.M., van Dongen, B.F., van der Aalst, W.M.P.: Towards Cross-Organizational Process Mining in Collections of Process Models and Their Executions. In: F. Daniel,

K. Barkaoui, S. Dustdar (eds.) Business Process Management Workshops - BPM 2011 International Workshops, Clermont-Ferrand, France, August 29, 2011, Revised Selected Papers, Part II, pp. 2–13. Springer (2011)

34. Buijs, J.C.A.M., van Dongen, B.F., van der Aalst, W.M.P.: A genetic algorithm for discovering process trees. In: IEEE Congress on Evolutionary Computation, pp. 1–8. IEEE (2012)

35. Buijs, J.C.A.M., van Dongen, B.F., van der Aalst, W.M.P.: On the Role of Fitness, Precision, Generalization and Simplicity in Process Discovery. In: R. Meersman, H. Panetto, T.S. Dillon, S. Rinderle-Ma, P. Dadam, X. Zhou, S. Pearson, A. Ferscha, S. Bergamaschi, I.F. Cruz (eds.) On the Move to Meaningful Internet Systems, OTM'12, *Lecture Notes in Computer Science*, vol. 7565, pp. 305–322. Springer (2012)

36. Buijs, J.C.A.M., Rosa, M.L., Reijers, H.A., van Dongen, B.F., van der Aalst, W.M.P.: Improving Business Process Models Using Observed Behavior. In: P. Cudré-Mauroux, P. Ceravolo, D. Gasevic (eds.) Data-Driven Process Discovery and Analysis, SIMPDA'12, *Lecture Notes in Business Information Processing*, vol. 162, pp. 44–59. Springer (2012)

37. Burattin, A., Sperduti, A.: PLG: A Framework for the Generation of Business Process Models and Their Execution Logs. In: M. zur Muehlen, J. Su (eds.) Business Process Management Workshops - BPM 2010 International Workshops and Education Track, Hoboken, NJ, USA, September 13-15, 2010, Revised Selected Papers, pp. 214–219. Springer (2010)

38. Carmona, J., Cortadella, J., Kishinevsky, M.: Genet: A Tool for the Synthesis and Mining of Petri Nets. In: Application of Concurrency to System Design, ACSD'09, pp. 181–185. IEEE Computer Society (2009)

39. Cormen, T.H., Leiserson, C.E., Rivest, R.L., Stein, C.: Introduction to Algorithms (3. ed.). MIT Press (2009)

40. van Dongen, B.F., Dijkman, R.M., Mendling, J.: Measuring Similarity between Business Process Models. In: Z. Bellahsene, M. Léonard (eds.) 20th International Conference on Advanced Information Systems Engineering, CAiSE'08, *Lecture Notes in Computer Science*, vol. 5074, pp. 450–464. Springer (2008)

41. van Dongen, B.F., Mendling, J., van der Aalst, W.M.P.: Structural Patterns for Soundness of Business Process Models. In: 10th IEEE International Enterprise Distributed Object Computing Conference, EDOC'06, pp. 116–128. IEEE Computer Society (2006)

42. Dumas, M., van der Aalst, W.M.P., ter Hofstede, A.H.M.: Process-Aware Information Systems: Bridging People and Software Through Process Technology. Wiley (2005)

43. Dumas, M., García-Bañuelos, L., Polyvyanyy, A.: Unraveling Unstructured Process Models. In: J. Mendling, M. Weidlich, M. Weske (eds.) Second International Workshop on Business Process Modeling Notation, BPMN'10, pp. 1–7. Springer (2010)

44. Fahland, D., van der Aalst, W.M.P.: Repairing Process Models to Reflect Reality. In: A.P. Barros, A. Gal, E. Kindler (eds.) Business Process Management, BPM'12, *Lecture Notes in Computer Science*, vol. 7481, pp. 229–245. Springer (2012)

45. Goedertier, S., Martens, D., Vanthienen, J., Baesens, B.: Robust Process Discovery with Artificial Negative Events. Journal of Machine Learning Research **10**, 1305–1340 (2009)

46. Greco, G., Guzzo, A., Pontieri, L., Saccà, D.: Discovering Expressive Process Models by Clustering Log Traces. IEEE Trans. Knowl. Data Eng. **18**(8), 1010–1027 (2006)

47. Hopcroft, J.E., Tarjan, R.E.: Dividing a Graph into Triconnected Components. SIAM J. Comput. **2**(3), 135–158 (1973)

48. Jensen, K., Kristensen, L.: Coloured Petri Nets. Springer Verlag (2009)

49. Karrer, B., Levina, E., Newman, M.E.J.: Robustness of community structure in networks. Phys. Rev. E **77**(4), 046,119 (2008). DOI 10.1103/PhysRevE.77.046119

50. de Leoni, M., van der Aalst, W.M.P.: Aligning Event Logs and Process Models for Multiperspective Conformance Checking: An Approach Based on Integer Linear Programming. In: F. Daniel, J. Wang, B. Weber (eds.) 11th International Conference on Business Process Management, BPM '13, *Lecture Notes in Computer Science*, vol. 8094, pp. 113–129. Springer (2013)

51. de Leoni, M., van der Aalst, W.M.P.: Data-aware process mining: discovering decisions in processes using alignments. In: S.Y. Shin, J.C. Maldonado (eds.) ACM Symposium on Applied Computing, SAC '13, pp. 1454–1461. ACM (2013)

52. Liggesmeyer, P.: Software-Qualität - testen, analysieren und verifizieren von Software. Spektrum Akadem. Verl. (2002)
53. Mannhardt, F., de Leoni, M., Reijers, H.v.d.A.W.M.P.: Balanced Multi-Perspective Checking of Process Conformance. Tech. rep., BPMcenter.org (2014). BPM Center Report BPM-14-07
54. de Medeiros, A.K.A.: Genetic Process Mining. Ph.D. thesis, Technische Universiteit Eindhoven, Eindhoven, The Netherlands (2006)
55. de Medeiros, A.K.A., van der Aalst, W.M.P., Weijters, A.J.M.M.: Quantifying process equivalence based on observed behavior. Data Knowl. Eng. **64**(1), 55–74 (2008)
56. de Medeiros, A.K.A., Weijters, A.J.M.M., van der Aalst, W.M.P.: Genetic process mining: an experimental evaluation. Data Min. Knowl. Discov. **14**(2), 245–304 (2007)
57. Munoz-Gama, J.: Algorithms for Process Conformance and Process Refinement. Master's thesis, Universitat Politecnica de Catalunya, Barcelona, Spain (2010)
58. Munoz-Gama, J.: Conformance Checking and Diagnosis in Process Mining. Ph.D. thesis, Universitat Politecnica de Catalunya, Barcelona, Spain (2014)
59. Munoz-Gama, J., Carmona, J.: A Fresh Look at Precision in Process Conformance. In: R. Hull, J. Mendling, S. Tai (eds.) 8th International Conference on Business Process Management, BPM'10, *Lecture Notes in Computer Science*, vol. 6336, pp. 211–226. Springer (2010)
60. Munoz-Gama, J., Carmona, J.: Enhancing precision in Process Conformance: Stability, confidence and severity. In: IEEE Symposium on Computational Intelligence and Data Mining, CIDM'11, pp. 184–191. IEEE (2011)
61. Munoz-Gama, J., Carmona, J.: A General Framework for Precision Checking. Special Issue on Intelligent and Innovative Computing in Business Process Management of International Journal of Innovative Computing, Information and Control **8**(7(B)), 5317–5339 (2012)
62. Munoz-Gama, J., Carmona, J., van der Aalst, W.M.: Single-Entry Single-Exit Decomposed Conformance Checking. Information Systems **46**, 102–122 (2014)
63. Munoz-Gama, J., Carmona, J., van der Aalst, W.M.P.: Conformance Checking in the Large: Partitioning and Topology. In: F. Daniel, J. Wang, B. Weber (eds.) 11th International Conference on Business Process Management, BPM'13, *Lecture Notes in Computer Science*, vol. 8094, pp. 130–145. Springer (2013). [Best Student Paper Award]
64. Munoz-Gama, J., Carmona, J., van der Aalst, W.M.P.: Hierarchical Conformance Checking of Process Models Based on Event Logs. In: J.M. Colom, J. Desel (eds.) 34th International Conference on Application and Theory of Petri Nets and Concurrency, PETRI NETS'13, *Lecture Notes in Computer Science*, vol. 7927, pp. 291–310. Springer (2013)
65. Murata, T.: Petri nets: Properties, analysis and applications. Proceedings of the IEEE **77**(4), 541–580 (1989)
66. OMG: Business Process Model and Notation (BPMN) Version 2.0. http://www.omg.org/spec/BPMN/2.0/ (2011)
67. OMG: Unified Modeling Language (UML) Version 2.0. www.omg.org/spec/UML/2.0/ (2011)
68. Petkovic, M., Prandi, D., Zannone, N.: Purpose Control: Did You Process the Data for the Intended Purpose? In: W. Jonker, M. Petkovic (eds.) 8th VLDB Workshop on Secure Data Management, SDM'11, *Lecture Notes in Computer Science*, vol. 6933, pp. 145–168. Springer (2011)
69. Polyvyanyy, A.: Structuring process models. Ph.D. thesis, University of Potsdam (2012)
70. Polyvyanyy, A., Vanhatalo, J., Völzer, H.: Simplified Computation and Generalization of the Refined Process Structure Tree. In: M. Bravetti, T. Bultan (eds.) 7th International Workshop on Web Services and Formal Methods, WSFM'10, *Lecture Notes in Computer Science*, vol. 6551, pp. 25–41. Springer (2010)
71. Ramadge, P.J., Wonham, W.M.: Supervisory control of a class of discrete event processes. SIAM Journal on Control and Optimization **25**(1), 206–230 (1987)
72. Reisig, W., Rozenberg, G.: Lectures on Petri nets I: Basic models - Advances in Petri nets, vol. 1491. Springer (1998)
73. Rozinat, A.: Process Mining: Conformance and Extension. Ph.D. thesis, Technische Universiteit Eindhoven, Eindhoven, The Netherlands (2010)

74. Rozinat, A., van der Aalst, W.M.P.: Conformance Testing: Measuring the Alignment Between Event Logs and Process Models. BETA Research School for Operations Management and Logistics (2005)

75. Rozinat, A., van der Aalst, W.M.P.: Conformance Testing: Measuring the Fit and Appropriateness of Event Logs and Process Models. In: C. Bussler, A. Haller (eds.) Business Process Management Workshops, vol. 3812, pp. 163–176 (2005)

76. Rozinat, A., van der Aalst, W.M.P.: Decision Mining in ProM. In: S. Dustdar, J.L. Fiadeiro, A.P. Sheth (eds.) Business Process Management, BPM'06, *Lecture Notes in Computer Science*, vol. 4102, pp. 420–425. Springer (2006)

77. Rozinat, A., van der Aalst, W.M.P.: Conformance checking of processes based on monitoring real behavior. Inf. Syst. **33**(1), 64–95 (2008)

78. Rozinat, A., de Medeiros, A.K.A., Günther, C.W., Weijters, A.J.M.M., van der Aalst, W.M.P.: The Need for a Process Mining Evaluation Framework in Research and Practice. In: A.H.M. ter Hofstede, B. Benatallah, H.Y. Paik (eds.) Business Process Management Workshops, *Lecture Notes in Computer Science*, vol. 4928, pp. 84–89. Springer (2007)

79. Rozinat, A., Veloso, M., van der Aalst, W.M.P.: Using Hidden markov models to evaluate the quality of discovered process models. Tech. Rep. BPM-08-10, BPMcenter.org (2008)

80. Sarbanes, P., Oxley, G., et al.: Sarbanes-Oxley Act of 2012 (2002)

81. Schrijver, A.: Theory of linear and integer programming. Wiley-Interscience series in discrete mathematics and optimization. Wiley (1999)

82. Silva, M., Teruel, E., Colom, J.M.: Linear Algebraic and Linear Programming Techniques for the Analysis of Place or Transition Net Systems. In: W. Reisig, G. Rozenberg (eds.) Petri Nets, *Lecture Notes in Computer Science*, vol. 1491, pp. 309–373. Springer (1996)

83. Smirnov, S., Weidlich, M., Mendling, J.: Business Process Model Abstraction Based on Behavioral Profiles. In: P.P. Maglio, M. Weske, J. Yang, M. Fantinato (eds.) 8th International Conference on Service-Oriented Computing, ICSOC, *Lecture Notes in Computer Science*, vol. 6470, pp. 1–16 (2010). DOI 10.1007/978-3-642-17358-5

84. Solé, M., Carmona, J.: Rbminer: A Tool for Discovering Petri Nets from Transition Systems. In: A. Bouajjani, W.N. Chin (eds.) Automated Technology for Verification and Analysis, ATVA'10, *Lecture Notes in Computer Science*, vol. 6252, pp. 396–402. Springer (2010)

85. vanden Broucke, S., De Weerdt, J., Vanthienen, J., Baesens, B.: On replaying process execution traces containing positive and negative events. Feb research report kbi 1311, KU Leuven (2013)

86. Vanhatalo, J., Völzer, H., Koehler, J.: The refined process structure tree. Data Knowl. Eng. **68**(9), 793–818 (2009)

87. de Weerdt, J., de Backer, M., Vanthienen, J., Baesens, B.: A Critical Evaluation Study of Model-log Metrics in Process Discovery. In: M. Muehlen, J. Su (eds.) Business Process Management Workshops, *Lecture Notes in Business Information Processing*, vol. 66, pp. 158–169. Springer Berlin Heidelberg (2011)

88. Weerdt, J.D., Backer, M.D., Vanthienen, J., Baesens, B.: A robust F-measure for evaluating discovered process models. In: IEEE Symposium on Computational Intelligence and Data Mining, CIDM'11, pp. 148–155. IEEE (2011)

89. Weerdt, J.D., Backer, M.D., Vanthienen, J., Baesens, B.: A multi-dimensional quality assessment of state-of-the-art process discovery algorithms using real-life event logs. Inf. Syst. **37**(7), 654–676 (2012)

90. Weerdt, J.D., vanden Broucke, S.K.L.M., Vanthienen, J., Baesens, B.: Active Trace Clustering for Improved Process Discovery. IEEE Trans. Knowl. Data Eng. **25**(12), 2708–2720 (2013)

91. Weidlich, M., Mendling, J., Weske, M.: Efficient Consistency Measurement Based on Behavioral Profiles of Process Models. IEEE Trans. Software Eng. **37**(3), 410–429 (2011). DOI 10.1109/TSE.2010.96

92. Weijters, A., van der Aalst, W.M.P., de Medeiros, A.K.A.: Process Mining with the Heuristics Miner-algorithm. In: BETA Working Paper Series, vol. WP 166 (2006)

93. van der Werf, J.M.E.M., van Dongen, B.F., Hurkens, C.A.J., Serebrenik, A.: Process Discovery using Integer Linear Programming. Fundam. Inform. **94**(3-4), 387–412 (2009)

94. Weske, M.: Business Process Management - Concepts, Languages, Architectures, 2nd Edition. Springer (2012)

Printed in the United States
By Bookmasters